Lecture Notes in Mathematics

Edited by A. Dold and B. Eckmann

811

Dag Normann

Recursion on the Countable Functionals

Springer-Verlag
Berlin Heidelberg New York 1980

Author

Dag Normann
Institute of Mathematics, The University of Oslo
Box 1053 Blindern Oslo 3
Norway

AMS Subject Classifications (1980): 03 D 65

ISBN 3-540-10019-9 Springer-Verlag Berlin Heidelberg New York
ISBN 0-387-10019-9 Springer-Verlag New York Heidelberg Berlin

Library of Congress Cataloging in Publication Data. Normann, Dag, 1947- Recursion on the countable functionals. (Lecture notes in mathematics; 811) Bibliography: p. Includes index. 1. Recursion theory. 2. Computable functions. I. Title. II. Series: Lecture notes in mathematics (Berlin); 811.
QA3.L28 no. 811. [QA96] 510s [511.3] 80-19391

© by Springer-Verlag Berlin Heidelberg 1980
Printed in Germany

Printing and binding: Beltz Offsetdruck, Hemsbach/Bergstr.
2141/3140-543210

Introduction

Generalized recursion theory is an area of mathematical logic which has been rapidly growing over the last twenty years, and it is now recognized as a dicipline of its own.

One reason for generalizing recursion theory is to find analogues to the natural numbers and the recursion theory on them, partly in order to use the intuition about computations on other domains and partly in order to generalize classical results. The consequence is that one quite often looks at theorems requiring a difficult combinatorial proof in a more general situation. Many of the proofs in generalized recursion theory are therefore true generalizations of classical proofs concerning recursion or metarecursion on the natural numbers, and much of their value is that they give a better understanding of the classical proofs.

Another reason for generalizing recursion theory is to look at other domains and 'true' algorithms on such domains. Then the motivation is not just to generalize but to find domains on which the notions of algorithm and computation make sense and to find what the algorithms and computations really are.

One attempt to extend recursion theory was done by Kleene during the fifties. He defined algorithms operating on functionals of arbitrary finite types and it is widely recognized that he gave an important analysis of the notion of a computation in a more general setting. Much of the work that has been done on Kleene's computation theory for higher types has been concerned with computations relative to certain functionals. The theory for recursion in the so-called normal functionals in particular has been a successful subbranch of the general theory. This theory is a generalization of the theory for hyperarithmetic sets and the research follows the patterns of generalized recursion theory described above. In particular computations will be infinite in a very strong sense.

Kleene isolated a subclass of his functionals of higher types, the countable functionals. It is a natural subclass if one wants to preserve some of the finiteness of ordinary recursion theory; a computation in a finite sequence of countable functionals may have an infinite computation tree but the value is decidable from a finite amount of information about the functionals involved. Kleene showed that they are closed under recursion.

Independently Kreisel, being interested in matters of constructivity, found a hierarchy of functionals suitable for arguments about constructive mathematics. He called them continuous functionals. It was clear that modulo unimportant differences these two classes of functionals were equivalent. This makes the countable or continuous functionals an interesting domain for recursion theory; computations are still finite in character (it has been shown that they cannot be extended preserving this property) and the hierarchy itself is as constructive as the real line.

Accepting this and accepting that recursion theory is worthwhile studying for its own sake one is motivated for the material in this book. This book is written out of an interest in the recursion theory of the countable functionals and related structures and it will mainly be concerned with recursion-theoretic problems.

After two introductory chapters we give in chapter 3 a structural analysis of the countable functionals, partly because this material has not been published elsewhere, partly because there is a structural understanding underlying most of the proofs to come later on. The rest of the book will contain various proofs of theorems concerning the recursion theory and the sample of proofs and results should be sufficient to introduce the reader to the main methods and problems in the area up to present research level. The book does not claim to contain all interesting results on the subject, just to enable the reader to understand other research papers.

The book is essentially self-contained. The reader is supposed to know ordinary recursion theory up to a good understanding of the basic notions and a knowledge of the basic terminology.

We will use a result of r.e. degree-theory but not the proof. In one of the proofs we will use a priority-argument and for the understanding of that proof it will be an advantage to have seen a priority-argument before. In Chapters 5 - 7 we need some elementary facts concerning the projective hierarchy but there is no advanced descriptive set theory involved.

My first contact with the continuous functionals took place in Oxford in Spring 1975. During my stay there I met Robin Gandy, Jan Bergstra, Martin Hyland and Stan Wainer and in the years to follow they inserted an interest for the subject in me. In particular discussions with Wainer and Bergstra concerning precise open problems made me work in the field.

The idea of writing a book on the subject grew out of a seminar I gave in Oslo in the autumn-term 1977. John V. Tucker suggested it

to me and Jens Erik Fenstad encouraged both of us. Tucker also gave
some valuable suggestions on the content of the book. Inspired by the
visit of Stan Wainer to Oslo in the spring-term '78 the material started
to take shape and the actual work with the manuscript took place autumn
'78 and early spring '79. The final version gives the status by Easter
'79, no later results gave been incorporated.

During my toiling with the manuscript my wife Svanhild has read
the various drafts and given valuable suggestions both concerning the
English and the way of presenting the material. Her assistance has
been most helpful. Stan Wainer was kind enough to read the final version of the manuscript and his comments induced some important changes.
I am sorry that I put this burden on these two but the book certainly
improved from it.

John Hartley later read parts of the manuscript and discovered
several minor errors in the text.

Finally I will express my gratitude towards R. Møller who performed
the skilled typing mostly while I was not around to decipher the hand-
written manuscript.

Oslo, January 1980
Dag Normann

CONTENTS

1. THE MAXIMAL TYPE STRUCTURE

1.1 Functionals of higher types

Ordinary recursion theory deals with computable operations on the natural numbers, or in some expositions (e.g. Shoenfield [43]), with computable operations on finite entities. But even then, when we want to compute on finite sequences, words or whatever we are interested in, we normally code our objects as natural numbers and translate the computation to a computation on natural numbers.

Not every interesting operator in mathematics deals with finite arguments and gives finite answers. A typical example is the partial operator

$$I(f,a,b) = \int_a^b f(x)dx$$

where a,b are reals and $f: \mathbb{R} \to \mathbb{R}$ is a function. Two of the arguments, a,b, are infinite sequences of finite entities (finite parts of the decimal expansion), while the first argument, f, is itself a function operating on infinite arguments and giving infinite answers. Does it make sense to ask whether the Rieman integral operator is computable or not?

Without doubt the numerical analysist will say that for decent f there are good algorithms that may be used by computers computing the integral up to any predecided accuracy, so there must be some notion of computability floating around.

The task of a mathematician is to take some phenomenon, analyze it, build a beautiful model for it and prove a lot of mathematically interesting properties of that model.

We will choose the Rieman integral and related operations.

When a computer computes an integral it is actually given a natural number n and is asked to give the n first decimals in the answer. So we deal in fact with the operator

$$I(f,a,b,n) = \int_a^b f(x)dx \quad \text{given with } n \text{ decimals.}$$

But now we have achieved something, the answers given by the operator are finite entities and may, as in ordinary recursion theory, be coded as natural numbers.

A real number a can be viewed as a function mapping a

natural number n onto the first n decimals of a, and the function f can also be viewed as operating on a real a and a natural number n, giving the n first decimals of $f(a)$.

Thus all the operators we consider can be regarded as operators with finite or infinite arguments, and giving natural numbers as values.

The discussion above should explain why we are interested in operators giving natural numbers as values and taking finite sequences of natural numbers and operators as arguments. This leads to the following definition of type-symbol σ and the type itself, $Tp(\sigma)$, denoted by it.

Definition 1.1

i   0 is a type-symbol denoting ω = the set of natural numbers.

ii   If $\sigma_1, \ldots, \sigma_n$ are type-symbols denoting $Tp(\sigma_1), \ldots, Tp(\sigma_n)$ resp., then $\sigma = (\sigma_1, \ldots, \sigma_n)$ is a type-symbol denoting

$$Tp(\sigma) = \text{The set of total functions} \quad \psi : Tp(\sigma_1) \times \cdots \times Tp(\sigma_n) \to \omega$$

Remark 1.2

In the literature one will often find the following alternative definition:

i   0 is a type-symbol denoting ω

ii   If σ and τ are type-symbols denoting $Tp(\sigma)$, $Tp(\tau)$ resp., then $(\sigma \to \tau)$ is a type-symbol denoting the set of total functions $\psi : Tp(\sigma) \to Tp(\tau)$.

The two definitions can be shown equivalent by iterating the following kind of transformations:

If $\psi : Tp(\sigma_1) \times \cdots \times Tp(\sigma_n) \to \omega$

replace it by

$$\psi' : Tp(\sigma_1) \to (Tp(\sigma_2) \times \cdots \times Tp(\sigma_n) \to \omega)$$

defined by

$$\psi'(\varphi_1) = \lambda(\varphi_2, \ldots, \varphi_n)\psi(\varphi_1, \ldots, \varphi_n)$$

where φ_1 varies over $Tp(\sigma_1)$, and for each φ_1, $\lambda(\varphi_2, \ldots, \varphi_n)\psi(\varphi_1, \ldots, \varphi_n)$ denotes the operator which to arguments $(\varphi_2, \ldots, \varphi_n)$ in $Tp(\sigma_2) \times \cdots \times Tp(\sigma_n)$ gives the value $\psi(\varphi_1, \varphi_2, \ldots, \varphi_n)$.

If $\phi : Tp(\sigma) \to Tp(\tau)$ where $\tau \neq 0$, then τ is of the form $\tau_1 \to \tau_2$, and we replace Φ by

$$\Phi':\text{Tp}(\sigma) \times \text{Tp}(\tau_1) \to \text{Tp}(\tau_2)$$

defined by

$$\Phi'(\psi,\varphi_1) = \Phi(\psi)(\varphi_1) .$$

Remark 1.3

The use of λ as above is of great notational importance. If $f(x_1,\ldots,x_n,y_1,\ldots,y_m)$ is a function, we will often be interested in the operator that to the arguments $y_1,\ldots,y_m$ gives the function

$$g(x_1,\ldots,x_n) = f(x_1,\ldots,x_n,y_1,\ldots,y_m) .$$

We will denote this function g by

$$\lambda(x_1,\ldots,x_n)f(x_1,\ldots,x_n,y_1,\ldots,y_m) .$$

We will not really be concerned with any of these two notions of functionals of higher types, the purification process will go on a bit further. Therefore we will not give detailed descriptions of the transformations indicated above. If one wants to give a precise definition, the following concept, telling "how far up" a type is, is of value.

Definition 1.4

To any type-symbol, σ, we associate a natural number, the level of σ or the type denoted by σ by

i   The level of the type-symbol 0 is 0

ii Let $\sigma_1,\ldots,\sigma_n$ be type-symbols with levels $k_1,\ldots,k_n$ resp. The level of $(\sigma_1,\ldots,\sigma_n)$ is then $1+\max\{k_1,\ldots,k_n\}$.

In order to test the understanding of this definition, we suggest the following exercise:

Prove: Let σ,τ be two type-symbols. Then σ and τ are of the same level if and only if $\text{Tp}(\sigma)$ and $\text{Tp}(\tau)$ have the same cardinality.

We are now going to define the objects that we really will be working with, the functionals of pure types. The pure types will be hand-picked representatives for each level of types, and we will denote them by natural numbers.

Definition 1.5

i   Let $\text{Tp}(0) = \omega$

<u>ii</u>   Let $Tp(k+1) = Tp(k)_\omega$

                        = The set of total functions $\psi : Tp(k) \to \omega$.

The aim of the last part of this section is to show that, inside
$<Tp(n)>_{n\in\omega}$ we have enough structure to simulate the broader types from
definition 1.1. Since we have not yet developed a computation theory,
we cannot prove that our codings are computable, but they are clearly
'effective' in some sense.

It is well known from ordinary recursion theory that there is a
recursive pairing function $<,> : \omega \times \omega \to \omega$ with recursive projection
functions $(\ )_1$ and $(\ )_2$ such that

$$(<n,m>)_1 = n \quad \text{and} \quad (<n,m>)_2 = m .$$

These definitions are directly lifted to $Tp(n+1)$ by

<u>Definition 1.6</u>

<u>a</u>   Let ψ_1, ψ_2 be elements in $Tp(n+1)$.
      Let $<\psi_1, \psi_2>$ be the element in $Tp(n+1)$ defined by

$$<\psi_1, \psi_2>(\varphi) = <\psi_1(\varphi), \psi_2(\varphi)> .$$

b)   Let $\psi \in Tp(n+1)$. Define $(\psi)_i$ $(i = 1,2)$ by

$$(\psi)_i(\varphi) = (\psi(\varphi))_i .$$

It follows that $(<\psi_1, \psi_2>)_1 = \psi_1$ and $(<\psi_1, \psi_2>)_2 = \psi_2$. Given the
pairing functions we can map any fixed number of elements $\psi_1, \dots, \psi_{n+1}$
into one single element by

$$<\psi_1, \dots, \psi_{n+1}>_{n+1} = <<\psi_1, \dots, \psi_n>_n, \psi_{n+1}>$$

where $<,>_2 = <,>$.

So any finite cartesian product of a fixed type may be identified
with the type itself.

In order to jump from one type to another, we need the push up and
push down operators given in the following definition.

<u>Definition 1.7</u>

<u>i</u>   If $n \in Tp(0)$, then $n^+ \in Tp(1)$ is defined by

$$n^+(m) = n$$

<u>ii</u>   If $f \in Tp(1)$, then $f^- \in Tp(0)$ is defined by

$$f^- = f(0)$$

<u>iii</u> If $\varphi \in Tp(n)$ and $n > 0$, then $\varphi^+ \in Tp(n+1)$ is defined by

$$\varphi^+(\psi) = \varphi(\psi^-)$$

<u>iv</u> If $\psi \in Tp(n+1)$ and $n > 0$, then $\psi^- \in Tp(n)$ is defined by

$$\psi^-(\varphi) = \psi(\varphi^+)$$

<u>v</u> If $n < m$, then $P_n^m : Tp(n) \to Tp(m)$ is defined by

$$P_n^m(\varphi) = \varphi^{+\cdots+} \quad \text{where the number of +'s is } m-n$$

If $n > m$, then $P_n^m : Tp(n) \to Tp(m)$ is defined by

$$P_n^m(\varphi) = \varphi^{-\cdots-} \quad \text{where the number of -'s is } n-m$$

If $n = m$, then P_n^m is the identity on $Tp(n)$.

<u>Remark 1.8</u>

φ^+ is called the <u>push-up</u> of φ and φ^- is called the <u>push-down</u>
of φ . When we push down we will loose some information, while we keep
all information by pushing up. We will later prove that P_n^m is computable. Now we will show that given the growth of cardinality with the
type, the P_n^m's are as faithful as possible.

<u>Lemma 1.9</u>

<u>a</u> If $n < m < k$ or $k < m < n$ then $P_n^k = P_m^k \circ P_n^m$.

<u>b</u> If $n < m$ and $\varphi \in Tp(n)$, then $P_m^n(P_n^m(\varphi)) = \varphi$.

<u>Proof</u>:

<u>a</u> is immediate from the definition. To prove <u>b</u>, it is sufficient to
prove that for all φ we have that $(\varphi^+)^- = \varphi$. We prove this by induction on the type of φ .

If $\varphi \in Tp(0)$, then φ is a natural number n , φ^+ is the constant function $f(m) = n$ and $f^- = f(0) = n$, so

$$(\varphi^+)^- = \varphi .$$

Now let $\varphi \in Tp(n+1)$ and assume that the claim holds for all functionals of lower types. Then

$$(\varphi^+)^-(\psi)$$

$$= (\varphi^+)(\psi^+) \quad \text{by definition of } (\varphi^+)^-$$

$$= \varphi((\psi^+)^-) \quad \text{by definition of} \quad \varphi^+$$

$$= \varphi(\psi) \qquad \text{by the induction hypothesis applied to} \quad (\psi^+)^- .$$

□

We challenge the reader to use the push-up maps and codings of sequences to give effective embeddings of each $Tp(\sigma)$ from Definition 1.1 into $Tp(k)$, where k is the level of σ.

From now on we will only work with functionals of pure types, so we do not require technical familiarity with arbitrary types. To us they only serve as an intermediate stage in constructing the pure types and showing that they cover in a coded way the phenomena we want to discuss in this book.

1.2 Kleene's Computations

By the introduction of oracles, ordinary recursion theory is easily relativized to functions $f: \omega \to \omega$, so the notion of a recursive or computable functional of type 2 is meaningfull.

Kleene [22] lifted the notion of computability to functionals of arbitrary types. Later in this section we will give the precise definition of Kleene-computations via the schemes S1 - S9. But we will first discuss some of the problems involved.

We are going to define a class of valid computations which takes sequences of functionals as arguments and gives natural numbers as values.

Clearly the successor-operator, the constant functions and the identity operator on the natural numbers are computable (S1-S3). Also, the composition of two valid computations must be a valid computation (S4), and the use of primitive recursion to define new computations must be permitted (S5).

If $\psi(\varphi_1, \ldots, \varphi_n)$ is a computable function and σ is a permutation on $\{1, \ldots, n\}$, then

$$\psi'(\varphi_1, \ldots, \varphi_n) = \psi(\varphi_{\sigma(1)}, \ldots, \varphi_{\sigma(n)})$$

must be regarded as computable (S6).

If f is of type 1 and $x \in \omega$, then $f(x)$ is clearly uniformly computable in f, x (S7).

Combining this with composition we see that if we have computed x by some algorithm, we compute $f(x)$ by using an oracle for f combined with the algorithm for x. Then we may use $f(x)$ further in some larger computation. Thus, regarding point-evaluation $f(x)$ on numbers as computable permits us to use functions in intermediate computations.

The most natural way to generalize S7 to higher types might seem to let the evaluation-operator $Ev(\psi,\varphi) = \psi(\varphi)$ be computable. If we would permit computations to take functionals as values we could use functional-application in intermediate computations. But life will be much easier if we can keep the natural numbers as the only possible values of our computations, and the need to evaluate ψ on an intermediately computed functional φ is our only reason for introducing functionals as values. Thus we introduce a new scheme which to a functional ψ and an algorithmic description of a functional φ gives us $\psi(\varphi)$ whenever this has meaning (S8).

Our algorithms will be indexed by numbers, and like most recursion theorists we believe that going from an index for an algorithm directly to the algorithm itself is effective (S9).

The exact notion of a Kleene-computation is defined by an inductive definition with nine clauses called schemes (S1-S9). To each scheme we associate a natural number, an index, which will give perfect coding of the actual algorithm. The index will be a coded sequence $<i,...,\sigma>$ where i is the number of the clause used, ... will contain special information (e.g. in the case of composition the indices for the two algorithms composed) and σ will be a coded sequence $<k_1,...,k_n>$ saying that arguments accepted by this algorithm should be of types $k_1,...,k_n$ in that order.

The expression $\underline{\{e\}(\vec{\psi})}$ means the algorithm with index e applied on $\vec{\psi}$, and if the algorithm works, $\{e\}(\vec{\psi})$ will also denote the value of the computation. This ambiguity is not greater than the one used in calculus, where $\sum_{n=1}^{\infty} a_n$ both means the sequence of finite partial sums and the limit whenever the limit exists.

We will now give the definition we are going to work with. It is not symbol by symbol as in Kleene [22], but we clearly define the same notion of computability.

Definition 1.10

S1. If $e = <1,\sigma>$, $x \in \omega$ and σ is the sequence (number) of the types of $(x,\varphi_1,...,\varphi_k)$, then

$$\{e\}(x,\varphi_1,...,\varphi_k) = x+1$$

S2. If $e = <2,q,\sigma>$, $q \in \omega$ and σ is the sequence of the types of $(\varphi_1,...,\varphi_k)$, then

$$\{e\}(\varphi_1,...,\varphi_k) = q$$

S3. If $e = \langle 3, \sigma \rangle$, $x \in \omega$ and σ is the sequence of the types of $(x, \varphi_1, \ldots, \varphi_k)$, then

$$\{e\}(x, \varphi_1, \ldots, \varphi_k) = x$$

S4. If $e = \langle 4, e_1, e_2, \sigma \rangle$ and σ is the sequence of the types of $(\varphi_1, \ldots, \varphi_k)$, then

$$\{e\}(\varphi_1, \ldots, \varphi_k) \simeq \{e_1\}(\{e_2\}(\varphi_1, \ldots, \varphi_k), \varphi_1, \ldots, \varphi_k)$$

where $a \simeq b$ means 'both a and b are defined and are equal, or they are both undefined'.

S5. If $e = \langle 5, e_1, e_2, \sigma \rangle$ and σ is the sequence of the types of $(x, \varphi_1, \ldots, \varphi_k)$, then

  $\underline{i}$ $\{e\}(x, \varphi_1, \ldots, \varphi_k) \simeq \{e_1\}(\varphi_1, \ldots, \varphi_k)$ if $x = 0$

  $\underline{ii}$ $\{e\}(x, \varphi_1, \ldots, \varphi_k) \simeq \{e_2\}(\{e\}(x-1, \varphi_1, \ldots, \varphi_k), \varphi_1, \ldots, \varphi_k)$ if $x > 0$

S6. If $e = \langle 6, e_1, \bar{\tau}, \sigma \rangle$, $\bar{\tau}$ codes a permutation τ of k elements and σ is the sequence of the types of $(\varphi_1, \ldots, \varphi_k)$, then

$$\{e\}(\varphi_1, \ldots, \varphi_k) \simeq \{e_1\}(\varphi_{\tau(1)}, \ldots, \varphi_{\tau(k)})$$

S7. If $e = \langle 7, \sigma \rangle$, $x \in \omega$, $f \in \mathrm{Tp}(1)$ and σ is the sequence of the types of $(x, f, \varphi_1, \ldots, \varphi_k)$, then

$$\{e\}(x, f, \varphi_1, \ldots, \varphi_k) = f(x)$$

S8. If $e = \langle 8, e_1, \sigma \rangle$ and σ is the sequence of the types of $(\varphi_1, \ldots, \varphi_k)$, then

$$\{e\}(\varphi_1, \ldots, \varphi_k) \simeq \varphi_1(\lambda \psi \{e_1\}(\psi, \varphi_1, \ldots, \varphi_k))$$

where the λ-notation is interpreted as in remark 1.3.

S9. If $e = \langle 9, t, \sigma \rangle$, $e_1 \in \omega$, $t \leq k$ and σ is the sequence of the types of $(e_1, \varphi_1, \ldots, \varphi_k)$, then

$$\{e\}(e_1, \varphi_1, \ldots, \varphi_k) \simeq \{e_1\}(\varphi_1, \ldots, \varphi_t)$$

Remark 1.11

   In S4 it is understood that $\{e\}(\varphi_1, \ldots, \varphi_k)$ is defined only if $\{e_2\}(\varphi_1, \ldots, \varphi_k)$ is defined with some value s and $\{e_1\}(s, \varphi_1, \ldots, \varphi_k)$ is defined.

   The same remark is valid for S5 and S6.

   In S8 it is understood that the variable ψ varies over functionals of two types less than the type of φ_1. The computation is defined if all computations $\{e_1\}(\psi,\varphi_1,\ldots,\varphi_k)$ are defined for ψ of the appropriate type, i.e. if $\lambda\psi\{e_1\}(\psi,\varphi_1,\ldots,\varphi_k)$ is a total functional of type one less than the type of φ_1.

The definition of the relation $\{e\}(\varphi_1,\ldots,\varphi_k) \simeq n$ may be regarded as an inductive definition Γ, as we will describe below. In discussing computations we will without mentioning it assume that the arguments are in the appropriate form for the index, and normally we discuss just a few cases covering the methods and ideas involved.

Although Γ is defined by 9 clauses corresponding to S1 - S9, we only give a sample of them here:

Definition 1.12

Define the operator $\Gamma(S)$ generating the <u>computation-tuples</u> of Kleene-recursion by

1. $<e,x,\varphi_1,\ldots,\varphi_k,x+1> \in \Gamma(S)$ when $e = <1,\sigma>$

2., 3. and 7. are analogous.

4. If $<e_2,\varphi_1,\ldots,\varphi_k,x>$ and $<e_1,x,\varphi_1,\ldots,\varphi_k,y>$ are both in S, then
   $<e,\varphi_1,\ldots,\varphi_k,y> \in \Gamma(S)$, where $e = <4,e_1,e_2,\sigma>$

5., 6. and 9. are analogous.

8. If $e = <8,e_1,\sigma>$ and for some functional ξ,

$$\forall\psi<e_1,\psi,\varphi_1,\ldots,\varphi_k,\xi(\psi)> \in S,$$

then $<e,\varphi_1,\ldots,\varphi_k,\varphi_1(\xi)> \in \Gamma(S)$.

A sequence is in $\Gamma(S)$ if it is in S or if it is in $\Gamma(S)$ by one of the nine conditions above.

Remark 1.13

Let

$$\Gamma^0 = \emptyset$$

$$\Gamma^\beta = \bigcup_{\gamma<\beta} \Gamma(\Gamma^\gamma)$$

and

$$\Gamma^\infty = \bigcup_{\alpha \in \text{Ordinals}} \Gamma^\alpha$$

We then see that $\{e\}(\varphi_1,\ldots,\varphi_k) \simeq n$ if and only if $<e,\varphi_1,\ldots,\varphi_k,n> \in \Gamma^\infty$.

Definitions 1.11 and 1.13 are various ways of formulating the same notion of computations, and the only thing we need from the definition in 1.12 is the feeling that the Kleene-computations are inductively defined. This leads us to the important fact that to any computation $\{e\}(\varphi_1,\ldots,\varphi_k) \simeq n$ there is associated an ordinal, the length of the computation, namely the stage in the induction at which the computation occurs. We will not stress this further, but give an example of how the notion of length may be used.

Lemma 1.14

Let $\{e\}(\varphi_1,\ldots,\varphi_k) \simeq n$ and $\{e\}(\varphi_1,\ldots,\varphi_k) \simeq m$ be two computations. Then $n = m$.

Proof:

We use induction on the length of the computation $\{e\}(\varphi_1,\ldots,\varphi_k) \simeq n$.

If the length is 1, we must have used S1-3 or S6. In all these cases we see by inspection that the value depends uniquely on the index and the arguments.

If the length is > 1 we must have used S4, S5, S8 or S9. We examine S4 and S8 more closely.

S4: $\{e\}(\varphi_1,\ldots,\varphi_k) \simeq \{e_1\}(\{e_2\}(\varphi_1,\ldots,\varphi_k),\varphi_1,\ldots,\varphi_k)$.

Now the computation $\{e_2\}(\varphi_1,\ldots,\varphi_k)$ has shorter length than the given one, and by the induction hypothesis $\{e_2\}(\varphi_1,\ldots,\varphi_k)$ has a unique value x. Also $\{e_1\}(x,\varphi_1,\ldots,\varphi_k)$ has shorter length than $\{e\}(\varphi_1,\ldots,\varphi_k)$ and has unique value y. But then $\{e\}(\varphi_1,\ldots,\varphi_k)$ has unique value y.

S8: $\{e\}(\varphi_1,\ldots,\varphi_k) = \varphi_1(\lambda\psi\{e_1\}(\psi,\varphi_1,\ldots,\varphi_k))$.

For each ψ, the computation $\{e_1\}(\psi,\varphi_1,\ldots,\varphi_k)$ has shorter length than $\{e\}(\varphi_1,\ldots,\varphi_k)$, so by the induction hypothesis

$$\xi = \lambda\psi\{e_1\}(\psi,\varphi_1,\ldots,\varphi_k)$$

is uniquely determined.
But $\varphi_1(\xi)$ has a fixed value, so $\{e\}(\varphi_1,\ldots,\varphi_k)$ is unique. □

We will now give some concepts needed for further discussions of computations. Some of the details are left for the reader.

Definition 1.15

a   If $\{e\}(\varphi_1,\ldots,\varphi_k) \simeq n$, we let $\|<e,\varphi_1,\ldots,\varphi_k>\|$, $\|\{e\}(\varphi_1,\ldots,\varphi_k)\|$ and $\|<e,\varphi_1,\ldots,\varphi_k,n>\|$ all denote the length of the computation, as defined in remark 1.13.

If $\{e\}(\varphi_1,\ldots,\varphi_k)$ has no value, we let $\|<e,\varphi_1,\ldots,\varphi_k>\| = \infty$.

<u>b</u>   Computations obtained by the use of S1-S3 or S7 are called <u>initial</u> <u>computations</u>.

<u>c</u>   By induction on the length of a computation we define the notion of <u>immediate sub-computation</u>.

<u>S1-3,7</u>: Initial computations do not have immediate sub-computations.

<u>S4</u>: $\{e\}(\varphi_1,\ldots,\varphi_k) \simeq \{e_1\}(\{e_2\}(\varphi_1,\ldots,\varphi_k),\varphi_1,\ldots,\varphi_k)$ has $\{e_2\}(\varphi_1,\ldots,\varphi_k)$ and $\{e_1\}(x,\varphi_1,\ldots,\varphi_k)$ as immediate sub-computations, where x is the value of $\{e_2\}(\varphi_1,\ldots,\varphi_k)$.

S5, 6 and 9 are treated similarly.

<u>S8</u>: $\{e\}(\varphi_1,\ldots,\varphi_k) = \varphi_1(\lambda\psi\{e_1\}(\psi,\varphi_1,\ldots,\varphi_k))$ has all $\{e_1\}(\psi,\varphi_1,\ldots,\varphi_k)$ as immediate sub-computations.

<u>d</u>   The <u>sub-computation</u> relation is obtained by transitivizing the immediate sub-computation relation. Thus the sub-computations of a computation are the immediate sub-computations together with the sub-computations of the immediate sub-computations.

<u>e</u>   The <u>computation tree</u> of a computation is the computation together with its sub-computations, all with the sub-computation relation.

<u>f</u> $\{e\}(\varphi_1,\ldots,\varphi_k)\!\downarrow$ means that $\{e\}(\varphi_1,\ldots,\varphi_k)$ has a value.

$\{e\}(\varphi_1,\ldots,\varphi_k)\!\uparrow$ means that $\{e\}(\varphi_1,\ldots,\varphi_k)$ has no value.

We sometimes use <u>terminates</u> instead of "has a value".

<u>Remark 1.16</u>

    If $\{e\}(\varphi_1,\ldots,\varphi_k)\!\downarrow$ then the computation tree is well-founded, i.e. there is no infinite descending chain of sub-computations, and the height of the computation tree is exactly the length of the computation.

    The following lemma is proved by an easy induction on the length of the computation. We leave the proof for the reader.

<u>Lemma 1.17</u>

<u>a</u>   If $\{e\}(\varphi_1,\ldots,\varphi_k)\!\downarrow$, then all arguments φ_i that are not natural numbers occur as arguments in all sub-computations of $\{e\}(\varphi_1,\ldots,\varphi_k)$.

<u>b</u>   If $\{e\}(\varphi_1,\ldots,\varphi_k)\!\downarrow$ and $n \geq 2$ is the maximum of the types of $\varphi_1,\ldots,\varphi_k$, then for all sub-computations $\{e'\}(\psi_1,\ldots,\psi_t)$, if the type of ψ_i is $\geq n-1$, then $\psi_i \in \{\varphi_1,\ldots,\varphi_k\}$.

<u>Remark 1.18</u>

    Lemma 1.17 gives us some control over the kind of arguments that may occur in a sub-computation of a given computation. This will be of importance when we later are going to investigate the complexity of the

computation-relation.

In this paragraph we will state without proof some general recursion-theoretic facts about Kleene-recursion. We omit the proofs since they have nothing to do with the special domains that we investigate. The results are so standard that most readers will accept them or work out their own proofs. For those who want a printed version, Fenstad [11] gives a general treatment of basic facts about computation theories.

Theorem 1.19   (The S_m^n-theorem)

Fix the types of the arguments $\varphi_1,\ldots,\varphi_n,\psi_1,\ldots,\psi_m$. Then there is a recursive functional S_m^n such that for all e , $\varphi_1,\ldots,\varphi_n$, $\psi_1,\ldots,\psi_m$ of the fixed types we have

$$\{e\}(\varphi_1,\ldots,\varphi_n,\psi_1,\ldots,\psi_m) \simeq \{S_m^n(e,\varphi_1,\ldots,\varphi_n)\}(\psi_1,\ldots,\psi_m) .$$

Moreover, an index for S_m^n may be obtained by a primitive recursive function from n,m and the sequences of types of $(\varphi_1,\ldots,\varphi_n)$ and $(\psi_1,\ldots,\psi_m)$.

Theorem 1.20   (The recursion theorem)

Let e_1 be an index for an algorithm taking a number e and a sequence $\varphi_1,\ldots,\varphi_k$ as arguments.

Then there is an index e' such that

$$\forall(\varphi_1,\ldots,\varphi_k)[\{e_1\}(e',\varphi_1,\ldots,\varphi_k) \simeq \{e'\}(\varphi_1,\ldots,\varphi_k)] .$$

Remark 1.21

The recursion theorem enables us to define a recursive function by refering to the index for it. Our standard use will be to construct something recursive by induction on the length of computations and during the construction assume that we already have succeeded on all sub-computations.

The functionals generated by S1-S8 are called the primitive recursive functionals of higher type. They are all total. This class should not be confused with Gödels impredicative primitive recursive functionals of higher type.

Lemma 1.22

a   The pairing-function of any type is primitive recursive, i.e. the function ξ defined by

$$\xi(\psi_1,\psi_2,\varphi) = \langle\psi_1,\psi_2\rangle(\varphi)$$

is primitive recursive.

<u>b</u>   The push-up operator is primitive recursive, i.e. the function ξ defined by

$$\xi(\varphi,\psi) = \varphi^+(\psi)$$

is primitive recursive.

<u>c</u>   The push-down operator is primitive recursive.

<u>Proof</u>:

<u>a</u>   The pairing-function $<,>$ on ω is known to be primitive recursive, so it is generated by S1-S6. By definition of ξ and the higher type pairing function we see

$$\xi(\psi_1,\psi_2,\varphi) = <\psi_1,\psi_2>(\varphi) = <\psi_1(\varphi),\psi_2(\varphi)>$$

In order to find an algorithm for ξ we must use S8 (or S7 if $\varphi \in Tp(0)$) and composition together with the algorithm for $<,>$ on ω. <u>b</u> and <u>c</u> are left for the reader.

<u>Remark 1.23</u>

<u>a</u>   It is proved by induction on the type that point-evaluation $\psi(\varphi)$ is primitive recursive in ψ, φ. If $x \in \omega$, $f:\omega \to \omega$, then the function

$$\xi_1(f,x) = f(x)$$

is primitive recursive by S7.

    Now, assume that ξ_n, evaluation of a type n object on a type n-1 object, is primitive recursive. Let $\psi \in Tp(n+1)$, $\varphi \in Tp(n)$. Then

$$\xi_{n+1}(\psi,\varphi) = \psi(\varphi) = \psi(\lambda\eta\varphi(n)) = \psi(\lambda\eta\xi_n(\varphi,n))$$

This fits into S8, so ξ_{n+1} is also primitive recursive.

<u>b</u>   In the proof of lemma 1.22 <u>a</u> we used composition at two places at a time. Formally, to do this we must use S4 twice, and S6 in between. But recursion-theory will easily get unnecessarily complicated if we require absolute stringency. Thus we will avoid trivial but tedious and obscuring manipulations with the schemes S1-S9.

<u>Definition 1.24</u>

<u>a</u>   An operator $\xi:Tp(k_1)\times\cdots\times Tp(k_n) \to \omega$ is called <u>computable</u> or <u>recursive</u> if there is an index e such that for all $(\varphi_1,\ldots,\varphi_n) \in$

$Tp(k_1) \times \cdots \times Tp(k_n)$ we have that

$$\{e\}(\varphi_1, \ldots, \varphi_n) \simeq \xi(\varphi_1, \ldots, \varphi_n)$$

b A subset X of $Tp(k_1) \times \cdots \times Tp(k_n)$ is called computable or recursive if the characteristic function of X is computable.

c A subset X of $Tp(k_1) \times \cdots \times Tp(k_n)$ is called semicomputable or semirecursive if there is an index e such that for all $(\varphi_1, \ldots, \varphi_n) \in Tp(k_1) \times \cdots \times Tp(k_n)$ we have that

$$(\varphi_1, \ldots, \varphi_n) \in X \longleftrightarrow \{e\}(\varphi_1, \ldots, \varphi_n) \simeq 0$$

d Let $\psi \in Tp(k)$. An operator $\xi: Tp(k_1) \times \cdots \times Tp(k_n) \to \omega$ is called computable in ψ , ψ-computable, recursive in ψ or ψ-recursive if there is an index e such that for all $(\varphi_1, \ldots, \varphi_n) \in Tp(k_1) \times \cdots \times Tp(k_n)$ we have that

$$\{e\}(\psi, \varphi_1, \ldots, \varphi_n) \simeq \xi(\varphi_1, \ldots, \varphi_n)$$

e Let $\psi \in Tp(k)$. A subset X of $Tp(k_1) \times \cdots \times Tp(k_n)$ is computable in ψ , ψ-computable, recursive in ψ or ψ-recursive if its characteristic function is ψ-computable.

f Let $\psi \in Tp(k)$. A subset X of $Tp(k_1) \times \cdots \times Tp(k_n)$ is semicomputable in ψ or semirecursive in ψ if there is an index e such that for all $(\varphi_1, \ldots, \varphi_n) \in Tp(k_1) \times \cdots \times Tp(k_n)$ we have that

$$(\varphi_1, \ldots, \varphi_n) \in X \longleftrightarrow \{e\}(\psi, \varphi_1, \ldots, \varphi_n) \simeq 0$$

g Let $\psi \in Tp(n)$, $k > 0$. By the k-section of ψ we mean

$$k\text{-sc}(\psi) = \{\varphi \in Tp(k); \varphi \text{ is } \psi\text{-computable}\}$$

By the k-envelope of ψ we mean

$$k\text{-en}(\psi) = \{X \subseteq Tp(k-1); X \text{ is semicomputable in } \psi\}$$

Remark 1.25

a In the literature one will find both computable and recursive used. We will use computable, since we in another context will have to distinguish between various notions of 'effective', and we will then use the term recursive in a different way.

b Sometimes the k-section of ψ will denote the subsets of $Tp(k-1)$ computable in ψ . Since we recursively may identify a functional with its graph, the two notions are equivalent. This identification is not valid if we deal with primitive recursion, because we need μ to compute ψ from its graph.

<u>c</u>   In the definition of semicomputability we could equivalently write $\{e\}(\varphi_1,\ldots,\varphi_n)\!\downarrow$ instead of $\{e\}(\varphi_1,\ldots,\varphi_n) \simeq 0$; we would get the same class of semicomputable sets.

In this paragraph we will illustrate one of the standard applications of the recursion theorem (Theorem 1.20). We first prove a simple lemma in detail, and we then prove the transitivity of the relation 'computable in' more in the style that we will normally use.

<u>Lemma 1.26</u>

There is a recursive map $\rho(e_0,i,j)$ such that if

$$\{e_0\}(\varphi_1,\ldots,\varphi_n)\!\downarrow \;,\quad \varphi_i = \varphi_j \quad \text{and} \quad i < j \leq n$$

then

$$\{\rho(e_0,i,j)\}(\varphi_1,\ldots,\varphi_{j-1},\varphi_{j+1},\ldots,\varphi_n) \simeq \{e_0\}(\varphi_1,\ldots,\varphi_n)$$

i.e. we may take φ_j away, using φ_i everywhere instead.

<u>Proof</u>:

We will define a recursive function $\nu(e,e_0,i,j)$, and by the recursion-theorem there will be an index e_1 such that for all e_0,i,j

$$\{e_1\}(e_0,i,j) \simeq \nu(e_1,e_0,i,j)$$

We are going to let $\rho(e_0,i,j) \simeq \{e_1\}(e_0,i,j)$

The construction of ν is by ten cases <u>o-ix</u>, where <u>i-ix</u> corresponds to the S1-S9 schemes.

<u>o</u> e_0 does not look like an index or $e_0 = <k,\ldots, \sigma>$, but either $\neg\,(1 \leq i < j \leq \mathrm{lh}(\sigma))$ or $\sigma_i \neq \sigma_j$ (where $\mathrm{lh}(\sigma)$ is the length of the sequence σ and σ_n is the n'th element of the sequence).

Then let $\nu(e,e_0,i,j) = 0$

In the remaining cases we assume that case <u>o</u> does not hold, and we let σ' be obtained from σ by removing element no. j .

<u>i</u> $e_0 = <1, \sigma>$.   Let $\nu(e,e_0,i,j) = <1, \sigma'>$

<u>ii</u> $e_0 = <2,q, \sigma>$.   Let $\nu(e,e_0,i,j) = <2,q, \sigma'>$

<u>iii</u> $e_0 = <3, \sigma>$.   Let $\nu(e,e_0,i,j) = <3, \sigma'>$

<u>iv</u> $e_0 = <4,e_2,e_3, \sigma>$

Let $\nu(e,e_0,i,j) = <4,\{e\}(e_2,i+1,j+1),\{e\}(e_3,i,j), \sigma'>$.

<u>v</u> $e_0 = <5,e_2,e_3, \sigma>$

Let $\nu(e,e_0,i,j) = <5,\{e\}(e_2,i,j),\{e\}(e_3,i+1,j+1), \sigma'>$

<u>vi</u> $e_0 = <6, e_2, \bar{\tau}, \sigma>$

If $\tau(i) < \tau(j)$, let $\nu(e, e_0, i, j) = <6, \{e\}(e_2, \tau(i), \tau(j)), \bar{\tau}', \sigma'>$ where τ' is obtained from τ by removing element no. j . If $\tau(i) > \tau(j)$, let $\nu(e, e_0, i, j) = <6, \{e\}(e_2, \tau(j), \tau(i)), \bar{\tau}', \sigma'>$ where τ' is obtained from τ by first interchanging elements i and j and then remove the element in position j .

<u>vii</u> $e_0 = <7, \sigma>$.   Let $\nu(e, e_0, i, j) = <7, \sigma'>$

<u>viii</u> $e_0 = <8, e_2, \sigma>$.   Let $\nu(e, e_0, i, j) = <8, \{e\}(e_2, i+1, j+1), \sigma'>$

<u>ix</u> $e_0 = <9, t, \sigma>$

We regard two subcases

<u>j > t</u>:   Let $\nu(e, e_0, u, j) = <9, t, \sigma'>$

<u>j ≤ t</u>:   Let $\nu(e, e_0, i, j) = <4, <9, t-1, \sigma'>, e_2, \sigma'>$ where e_2 is the index such that for all x

$$\{e_2\}(x) \simeq \{e\}(x, i-1, j-1)$$

e_2 is found uniformly in e, i, j using the S_m^n-theorem (Theorem 1.19).

This ends the construction of ν . By the recursion theorem, pick e_1 and define ρ as above.

We will now prove by induction on the length of the computation $\{e_0\}(\varphi_1, \ldots, \varphi_n)$ that ρ satisfies the lemma.

If $\{e_0\}(\varphi_1, \ldots, \varphi_n)$ is an initial computation, this is trivial.

We will content ourselves with proving the induction step for S4, S8 and S9 as the most interesting cases. So, we will assume that the lemma holds for all sub-computations.

<u>S4</u> $\{e_0\}(\varphi_1, \ldots, \varphi_n) \simeq \{e_2\}(\{e_3\}(\varphi_1, \ldots, \varphi_n), \varphi_1, \ldots, \varphi_n)$.

Then $\{\rho(e_0, i, j)\}(\varphi_1, \ldots, \varphi_{j-1}, \varphi_{j+1}, \ldots, \varphi_n)$

$\overset{D}{\simeq} \{\rho(e_2, i+1, j+1)\}(\{\rho(e_3, i, j)\}(\varphi_1, \ldots, \varphi_{j-1}, \varphi_{j+1}, \ldots, \varphi_n)$,

$\qquad \varphi_1, \ldots, \varphi_{j-1}, \varphi_{j+1}, \ldots, \varphi_n)$

$\overset{IH}{\simeq} \{\rho(e_2, i+1, j+1)\}(\{e_3\}(\varphi_1, \ldots, \varphi_n), \varphi_1, \ldots, \varphi_{j-1}, \varphi_{j+1}, \ldots, \varphi_n)$

$\overset{IH}{\simeq} \{e_2\}(\{e_3\}(\varphi_1, \ldots, \varphi_n), \varphi_1, \ldots, \varphi_n)$

$\simeq \{e_0\}(\varphi_1, \ldots, \varphi_n)$

<u>S8</u> $\{\rho(e_0, i, j)\}(\varphi_1, \ldots, \varphi_{j-1}, \varphi_{j+1}, \ldots, \varphi_n)$

$\overset{D}{\simeq} \varphi_1(\lambda\beta\{\rho(e_2, i+1, j+1\}(\beta, \varphi_1, \ldots, \varphi_{j-1}, \varphi_{j+1}, \ldots, \varphi_n))$

$\overset{IH}{\simeq} \varphi_1(\lambda\beta\{e_2\}(\beta, \varphi_1, \ldots, \varphi_n)) \simeq \{e_0\}(\varphi_1, \ldots, \varphi_n)$

<u>S9</u>   We must split the proof in two. Note that φ_1 is a number x.

<u>j > t:</u>

$$\{\rho(e_0,i,j)\}(x,\varphi_2,\ldots,\varphi_{j-1},\varphi_{j+1},\ldots,\varphi_n) \overset{D}{\simeq} \{x\}(\varphi_2,\ldots,\varphi_t)$$
$$\simeq \{e_0\}(x,\varphi_2,\ldots,\varphi_n)$$

<u>j ≤ t:</u>

$$\{\rho(e_0,i,j)\}(x,\varphi_2,\ldots,\varphi_{j-1},\varphi_{j+1},\ldots,\varphi_n)$$
$$\overset{D}{\simeq} \{\rho(x,i-1,j-1)\}(\varphi_2,\ldots,\varphi_{j-1},\varphi_{j+1},\ldots,\varphi_t)$$
$$\overset{IH}{\simeq} \{x\}(\varphi_2,\ldots,\varphi_t)$$
$$\simeq \{e_0\}(x,\varphi_2,\ldots,\varphi_n)$$

This ends the proof of the lemma.

□

Remark 1.27

This is a typical <u>reindexing</u> argument. We have a computation in one list of arguments and want to simulate it as a computation in another list of arguments. Normally the main difficulty in such proofs are found in connection with S8, and we will then concentrate on that.

In such reindexing arguments the reindexing function ρ normally turns out to be primitive recursive. In order to prove it one must in each case use the recursion theorem for primitive recursion.

Our next result will be proved more in the style we will use later, where we omit the trivial cases and formal definitions.

Theorem 1.28

There is a recursive function ρ such that whenever

$$\{e\}(\alpha_1,\ldots,\alpha_{n_1},\lambda\varphi\{d\}(\varphi,\beta_1,\ldots,\beta_{n_2}),\gamma_1,\ldots,\gamma_{n_3}) \simeq k$$

then

$$\{\rho(e,d,n_1,n_2,n_3)\}(\alpha_1,\ldots,\alpha_{n_1},\beta_1,\ldots,\beta_{n_2},\gamma_1,\ldots,\gamma_{n_3}) \simeq k$$

where the α's, β's and γ's are functionals.

<u>Proof:</u> We will use the recursion theorem to define ρ . The only difficulty is in connection with S8, when $n_1 = 0$, i.e. when we have that

$$\{e\}(\lambda\varphi\{d\}(\varphi,\beta_1,\ldots,\beta_{n_2}),\gamma_1,\ldots,\gamma_{n_3})$$
$$\simeq (\lambda\varphi\{d\}(\varphi,\beta_1,\ldots,\beta_{n_2})(\lambda\beta\{e_1\}(\beta,\lambda\varphi\{d\}(\varphi,\beta_1,\ldots,\beta_{n_2}),\gamma_1,\ldots,\gamma_{n_3})))$$
$$\simeq \{d\}(\lambda\beta\{e_1\}(\beta,\lambda\varphi\{d\}(\varphi,\beta_1,\ldots,\beta_{n_2}),\gamma_1,\ldots,\gamma_{n_3}),\beta_1,\ldots,\beta_{n_2})$$

The proof is by a double induction, primarily by induction on k = the type of φ , and, for a fixed k , by induction on the length of the computation

$$\{e\}(\alpha_1,\ldots,\alpha_{n_1},\lambda\varphi\{d\}(\varphi,\beta_1,\ldots,\beta_{n_2}),\gamma_1,\ldots,\gamma_{n_3}) \simeq k$$

Let us now assume that we are in the special case above and that the induction hypothesis holds for the ρ we are constructing.

In the computation

$$\{d\}(\lambda\beta\{e_1\}(\beta,\lambda\varphi\{d\}(\varphi,\beta_1,\ldots,\beta_{n_2}),\gamma_1,\ldots,\gamma_{n_3}),\beta_1,\ldots,\beta_{n_2})$$

we see that for each β

$$\{e_1\}(\beta,\lambda\varphi\{d\}(\varphi,\beta_1,\ldots,\beta_{n_2}),\gamma_1,\ldots,\gamma_{n_3})$$

is a subcomputation of the given computation, so by the induction hypothesis it will have the same value as

$$\{\rho(e_1,d,1,n_2,n_3)\}(\beta_0\beta_1,\ldots,\beta_{n_2},\gamma_1,\ldots,\gamma_{n_3})$$

so the given computation has the same value as

$$\{d\}(\lambda\beta\{\rho(e_1,d,1,n_2,n_3)\}(\beta,\beta_1,\ldots,\beta_{n_2},\gamma_1,\ldots,\gamma_{n_3}),\beta_1,\ldots,\beta_{n_2})$$

By a closer inspection we see that the type of β must be $k-1$, so by the induction hypothesis, the value will be

$$\{\rho(d,\rho(e_1,d,1,n_2,n_3),0,n_2+n_3,n_2)\}(\beta_1,\ldots,\beta_{n_2},\gamma_1,\ldots,\gamma_{n_3},\beta_1,\ldots,\beta_{n_2})$$

By lemma 1.26 we can avoid the use of duplicate arguments by recursively choosing a new index. So we let $\rho(e,d,0,n_2,n_3) =$
$\nu(\rho(d,\rho(e_1,d,1,n_2,n_3),0,n_2+n_3,n_2))$ where ν is recursive with the property that for all e', n_2, n_3, β's and γ's

$$\{e'\}(\beta_1,\ldots,\beta_{n_2},\gamma_1,\ldots,\gamma_{n_3},\beta_1,\ldots,\beta_{n_2}) \simeq \{\nu(e')\}(\beta_1,\ldots,\beta_{n_2},\gamma_1,\ldots,\gamma_{n_3})$$

□

This theorem has an important corollary, the relation 'computable in' between functionals of pure types is transitive.

Corollary 1.29
     Let φ, ψ and ξ be functionals of pure types. If φ is computable in ψ and ψ is computable in ξ , then φ is computable in ξ .

Proof: Let

$$\varphi = \lambda\alpha\{e\}(\alpha,\psi) \quad \text{and} \quad \psi = \lambda\beta\{e'\}(\beta,\xi)$$

Then

$$\varphi = \lambda\alpha\{e\}(\alpha,\lambda\beta\{e'\}(\beta,\xi))$$

$$= \lambda\alpha\{\rho(e,e',1.1.0)\}(\alpha,\xi)$$

where ρ is as in Theorem 1.28.

But this shows that φ is recursive in ξ .

□

In this book we will not get more involved with general properties of Kleene recursion. We think that the original papers of Kleene [22] serve as a good introduction to this theory, and we advice any reader with an appetite for this material to consult these papers.

1.3 A survey of recursion in normal functionals

Since Kleene's definition in [22], the concept of computations in higher type functionals has been deeply investigated. The functionals were soon divided in two classes, the normal functionals and the non-normal functionals. The normal functionals have strong computation power, giving nice regularity results, and this well-behavior was the main reason for calling them normal. Among the non-normal functionals, the continuous or countable functionals form an important subclass.

The literature on various aspects of computations in normal funct-ionals is rich, Hinman [18], Fenstad [11], Moldestad [27] and Sacks [42] all give good introductions to this theory, and the reader who is inter-ested in this part of higher recursion theory may consult one of these books. Here we will confine ourselves to giving a survey without proofs of the main aspects of computation in normal functionals.

The functional ^2E is the functional of type 2 defined as follows

$$^2E(f) = \begin{cases} 0 & \text{if } \forall x\, f(x) = 0 \\ 1 & \text{if } \exists x\, f(x) \neq 0 \end{cases}$$

A functional F of type 2 is called <u>normal</u> if ^2E is recursive in F .

The following results about ^2E are mainly due to Kleene.

Theorem 1.30

<u>a</u> $1\text{-sc}(^2E) = \Delta_1^1 = Tp(1) \cap L_{\omega_1^{CK}}$

where ω_1^{CK} is the 1st nonrecursive ordinal and L_α is Gödel's constructible hierarchy up to level α .

<u>b</u> $1\text{-en}(^2E) = \Pi_1^1$

<u>c</u> $X \subseteq Tp(1)$ is Borel if and only if there is an $f \in Tp(1)$ such that X is computable in $f, {}^2E$.

<u>d</u> $X \subseteq Tp(1)$ is co-analytic if and only if there is an $f \in Tp(1)$ such that X is semicomputable in $f, {}^2E$.

□

The following results about recursion in an arbitrary normal type-2 functional are mainly due to R.O. Gandy.

Theorem 1.31

<u>a</u>   Let F be normal and of type 2. Then there is an F-computable partial function p such that

if $\{e_1\}(F,f_1,\ldots,f_n)\!\downarrow$ or $\{e_2\}(F,g_1,\ldots,g_k)\!\downarrow$

where $f_i, g_j \in Tp(0) \cup Tp(1)$

then

$$p(<e_1,f_1,\ldots,f_n>,<e_2,g_1,\ldots,g_k>)\!\downarrow$$

and

$$p(<e_1,f_1,\ldots,f_n>,<e_2,g_1,\ldots,g_k>) \simeq \begin{cases} 0 & \text{if } \|<e_1,f_1,\ldots,f_n>\| \leq \|<e_2,g_1,\ldots,g_k>\| \\ 1 & \text{if } \|<e_1,f_1,\ldots,f_n>\| > \|<e_2,g_1,\ldots,g_k>\| \end{cases}$$

<u>b</u>   Let F be normal and of type 2. Then there is an F-computable partial function s such that if

$\exists n\{e\}(F,n,f_1,\ldots,f_k)\!\downarrow$ (where $f_1,\ldots,f_k \in Tp(0)\cup Tp(1)$)

then $s(<e,f_1,\ldots,f_k>)\!\downarrow$

and $\{e\}(F,s(<e,f_1,\ldots,f_k>),f_1,\ldots,f_k)\!\downarrow$

<u>c</u>   Let F be normal and of type 2 , $X \subseteq Tp(1)$.
Then

$$X \in 1\text{-sc}(F) \longleftrightarrow X \in 1\text{-en}(F) \wedge (Tp(1)\smallsetminus X) \in 1\text{-en}(F)$$

□

Definition 1.32

<u>a</u>   Let x be a set, A a relation on ω . We call A <u>a code</u> for x if A is isomorphic to the structure

<Transitive closure of $x, \in$ >

<u>b</u>   A class M of sets is called <u>locally countable</u> if all sets $x \in M$ have codes in M .

<u>c</u> If $\mathcal{O}$ is a family of subsets of ω , we let the <u>structure of $\mathcal{O}$</u> be

$$\text{Str}(\mathcal{O}) = \{x; x \text{ has a code in } \mathcal{O}\}$$

<u>d</u> A class M of sets is called <u>an abstract 1-section</u> if M is ad-
missible, countable, locally countable and satisfies Δ_0-dependent
choice.

The following characterization of 1-sc(F) for normal F was given
by G.E. Sacks [40].

<u>Theorem 1.33</u>

<u>a</u> If F is a normal functional of type 2 and $\mathcal{O}$ = 1-sc(F) , then
Str($\mathcal{O}$) is an abstract 1-section.

<u>b</u> If M is an abstract 1-section, then there is a normal functional
F such that

$$M \cap \text{Tp}(1) = 1\text{-sc}(F)$$

$\square$

There are higher type versions of ^2E defined as follows

$$^{k+2}E(\psi) = \begin{cases} 0 & \text{if } \forall\varphi \in \text{Tp}(k) \ \psi(\varphi) = 0 \\ 1 & \text{if } \exists\varphi \in \text{Tp}(k) \ \psi(\varphi) \neq 0 \end{cases}$$

A functional F of type k+2 is called <u>normal</u> if ^{k+2}E is computable
in F .

There are results analogous to Theorems 1.30 and 1.31 for normal
functionals of type > 2 . In order to state the more general version of
Theorem 1.30, a so called companion theory is developed. This is surv-
eyed e.g. in Fenstad [11].

The most striking property of recursion in a normal functional of
type k+2 , where $k \geq 1$, is the difference between the <u>individuals</u>, i.e.
Tp(k) , and the <u>sub-individuals</u>, i.e. Tp(k-1) . Moschovakis [31] proved
that if $F \in \text{Tp}(k+2)$ is normal and $k \geq 1$, then k+1-en(F) is not closed
under $\exists\varphi \in \text{Tp}(k)$. MacQueen [26] generalized this to all functionals.
On the other hand Harrington and MacQueen [16] proved that any nonempty
F-semicomputable subset of the subindividuals includes a nonempty F-
computable subset, and thus

$$k+1\text{-en}(F) \text{ is closed under } \exists\varphi \in \text{Tp}(k-1)$$

For further reading on recursion in normal higher type functionals
we refer to Moldestad [27]. Let us just quote two more results.

Theorem 1.34

a   The plus-one-theorem (Sacks [40] and [41]).

Let $k \geq 1$ and let H be a normal functional of type $\geq k+1$. Then there is a normal functional F of type $k+1$ such that

$$k\text{-sc}(F) = k\text{-sc}(H)$$

b   The plus-two-theorem Harrington [15])

Let $k \geq 1$, and let H be a normal functional of type $\geq k+2$. Then there is a normal functional F of type $k+2$ such that

$$k\text{-en}(F) = k\text{-en}(H)$$

□

Remark 1.35

We will later see that normality is essential in this theorem.

Problem 1.36

The following problem was suggested by Martin Hyland and is still unsolved:

If F is a functional of type > 2 and ^2E is computable in F, will then $\text{Str}(1\text{-sc}(F))$ be an abstract 1-section in the sense of definition 1.32 ?

Recursion in a normal functional has deep connections to admissibility theory and similar set-theoretic definability theory. This connection is established through the theory of set recursion (see Normann [33] for an introduction). In analogy with S1 - S9 we define a notion of recursion on the universe of sets, using the schemes for the rudimentary functions and adding a variety of S9.

The resulting theory is a generalization of recursion in a normal functional in a very direct way, for any set x the functional

$$^xE(f) = \begin{cases} 0 & \text{if } \forall y \in x \ f(y) = 0 \\ 1 & \text{if } \exists y \in x \ f(y) \neq 0 \wedge f \upharpoonright x \text{ is total} \end{cases}$$

is uniformly recursive in x.

In non-normal recursion theory we cannot translate the problems to definability theory in the same way. This may be the reason why this part of recursion in higher types has been slower in developing.

2. THE COUNTABLE FUNCTIONALS

2.1 Type-structures

In section 1.1 we defined the hierarchy of functionals of higher types, $\langle Tp(n) \rangle_{n \in \omega}$. In this section we will investigate a more general class of hierarchies of functionals, the so called type-structures. Later, much of our investigation will be concerned with one spesific example, the hierarchy of countable functionals, which we will define in the next section. The hierarchy $\langle Tp(n) \rangle_{n \in \omega}$ will from now on be called the <u>maximal type-structure</u>.

Definition 2.1

Let $\langle A_n \rangle_{n \in \omega}$ be a family of objects. $\langle A_n \rangle_{n \in \omega}$ is called a <u>type-structure if</u>

<u>i</u> $A_0 = \omega$

<u>ii</u> For all n

$$A_{n+1} \subseteq A_n \to \omega, \text{ i.e. } \psi \in A_{n+1} \to \psi \text{ is a total map of } A_n \text{ into } \omega.$$

The notion of Kleene-computations is meaningful for all type-structures. The only scheme we have to treat carefully is S8 :

$$\{e\}(\vec{\phi}) = \varphi_1(\lambda\psi\{e_1\}(\psi,\vec{\phi}))$$

Here we will assume that

$$\lambda\psi \in A_n\{e_1\}(\psi,\vec{\phi})$$

is an element of A_{n+1}, and that $\varphi_1 \in A_{n+2}$, before we accept the computation

$$\varphi_1(\lambda\psi\{e_1\}(\psi,\vec{\phi}))$$

This naturally leads to the following definition.

Definition 2.2

Let $\langle A_n \rangle_{n \in \omega}$ be a type-structure. We say that $\langle A_n \rangle_{n \in \omega}$ is <u>closed</u> <u>under Kleene-computation</u>, or just <u>closed under computation</u> if, whenever $\vec{\phi}$ is a sequence from $\langle A_n \rangle_{n \in \omega}$ and

$$\lambda\psi \in A_n\{e\}(\psi,\vec{\phi})$$

is total, then

$$\lambda\psi \in A_n\{e\}(\psi,\vec{\phi}) \in A_{n+1}$$

The general notion of a type-structure has been used with some success in Bergstra [1], Moldestad-Normann [28] and Moldestad [27]. We will develop some of the theory and indicate the main applications. First we will see how any type-structure may be "imbedded" in the maximal type-structure. Our first result was due to J. Bergstra [1].

Definition 2.3

Let $<A_n>_{n \in \omega}$. By induction on n we define for each $\varphi \in A_n$ a class $[\varphi]$ in $Tp(n)$ as follows.

i   If $m \in \omega$ let $[m] = \{m\}$

ii If $f \in A_1$ let $[f] = \{f\}$

iii If $\psi \in A_{n+1}$ let

$$[\psi] = \{\Psi \in Tp(n+1); \forall \varphi \in A_n \ \forall \Phi \in [\varphi](\psi(\varphi) = \Psi(\Phi))\}$$

Theorem 2.4

Let $A = <A_n>_{n \in \omega}$ be a type-structure closed under computation. Let $\psi_1, \ldots, \psi_k$ be functionals in A and let $\Psi_1, \ldots, \Psi_k$ be elements in $[\psi_1], \ldots, [\psi_k]$ resp.
Assume that

$$\{e\}(\Psi_1, \ldots, \Psi_k) = s$$

Then

$$\{e\}_A(\psi_1, \ldots, \psi_k) = s$$

where $\{e\}_A$ means that we interpret the computation in A .

Proof: The proof is by induction on the length of the computation

$$\{e\}(\Psi_1, \ldots, \Psi_k)$$

If the computation is initial, or actually anything but an application of S8, the argument is trivial. So assume

$$\{e\}(\Psi_1, \ldots, \Psi_k) = \Psi_1(\lambda\Phi\{e_1\}(\Phi, \Psi_1, \ldots, \Psi_k))$$

where Φ varies over $Tp(n)$.

Let $\varphi \in A_n$ be arbitrary, $\Phi \in [\varphi]$. By the induction hypothesis we have that

$$\{e_1\}_A(\varphi, \psi_1, \ldots, \psi_k) = \{e_1\}(\Phi, \Psi_1, \ldots, \Psi_k)$$

Let $\xi = \lambda\varphi\{e_1\}_A(\varphi, \psi_1, \ldots, \psi_k), \Xi = \lambda\Phi\{e_1\}(\Phi, \Psi_1, \ldots, \Psi_k)$
The argument above shows that $\Xi \in [\xi]$, so $\Psi_1(\Xi) = \psi_1(\xi)$.

But this means that

$$\{e\}_A(\psi_1,\ldots,\psi_k) = \{e\}(\psi_1,\ldots,\psi_k)$$

which is what we wanted to prove.

□

Corollary 2.5

Let $\langle A_n \rangle_{n\in\omega}$ be a type-structure closed under computations, and let $\varphi \rightsquigarrow [\varphi]$ be as defined in 2.3.

Let $\varphi \in A_n$, $\Phi \in [\varphi]$.

Then $1\text{-sc}(\Phi) \subseteq 1\text{-sc}_A(\varphi)$

where $1\text{-sc}_A(\varphi)$ is defined by relativizing the definition of a 1-section to A.

Proof: Let $f \in 1\text{-sc}(\Phi)$. Then f is computable in Φ, so there is an index e such that for all $n \in \omega$

$$f(n) = \{e\}(\Phi,n)$$

By Theorem 2.4 we then see that $f \in 1\text{-sc}_A(\varphi)$.

□

We will later want to construct pathological 1-sections by constructing a functional φ in some strange type-structure and then find a functional Φ in the maximal type-structure with the same 1-section. The next theorem, due to J. Bergstra [1], gives a method for constructing Φ from φ. In general Φ will not be an element of $[\varphi]$, but of $[\varphi']$ where φ' is some functional primitive recursively equivalent to φ.

Theorem 2.6

Let $A = \langle A_n \rangle_{n\in\omega}$ be a type-structure closed under computations. Let $\varphi \in A_n$. Then there is a functional $\Phi \in Tp(n)$ such that

$$1\text{-sc}_A(\varphi) = 1\text{-sc}(\Phi)$$

Proof: Let Ψ_k be the characteristic function of the [ ]-image of A_k, i.e.

$$\Psi_k(\Psi) = \begin{cases} 0 & \text{if } \exists \psi \in A_k \ (\Psi \in [\psi]) \\ 1 & \text{otherwise} \end{cases}$$

where [ ] is as in definition 2.3.

Let $^{k+1}0$ be the functional in A_{k+1} that is constant zero. Clearly $\Psi_k \in [^{k+1}0]$.

Let $\Phi' \in [\varphi]$ and let $\Phi = \langle \Phi', \Psi_1, \ldots, \Psi_{n-2} \rangle$. $\Phi \in [\langle \varphi, ^20, \ldots, ^{n-1}0 \rangle]$, so by corollary 2.5 we have that

$$1-sc(\phi) \subseteq 1-sc_A(<\varphi, {}^2 0, \ldots, {}^{n-1} 0>) = 1-sc_A(\varphi)$$

since ${}^2 0, \ldots, {}^{n-1} 0$ are computable in A.

<u>Claim</u>

There is a recursive map ρ such that for all e, $\varphi_1, \ldots, \varphi_k \in A_0 \cup \ldots \cup A_n$ and for all $\phi_1 \in [\varphi_1], \ldots, \phi_k \in [\varphi_k]$ we have that

$$\{e\}_A(\varphi_1, \ldots, \varphi_k) \simeq \{\rho(e)\}(\phi_1, \ldots, \phi_k, \Psi_1, \ldots, \Psi_{n-2})$$

Let us first see why the claim proves the theorem. We want to prove that $1-sc_A(\varphi) \subseteq 1-sc(\phi)$, so let $f \in 1-sc_A(\varphi)$, and choose e such that for all m

$$f(m) = \{e\}(m, \varphi)$$

By the claim then

$$f(m) = \{\rho(e)\}(m, \phi', \Psi_1, \ldots, \Psi_{n-2})$$

for all m, so f is computable in ϕ via index $\rho(e)$.

<u>Proof of claim</u>: We are following the pattern of a reindexing argument, see the proof of Lemma 1.26 as an example.

The construction of ρ and the proof that this works is by a simultaneous induction on the length of the computation

$$\{e\}_A(\varphi_1, \ldots, \varphi_k)$$

The only troublesome point is again when e is obtained by use of S8 :

$$\{e\}_A(\varphi_1, \ldots, \varphi_k) \simeq \varphi_1(\lambda\varphi \in A_m \{e_1\}_A(\varphi, \varphi_1, \ldots, \varphi_k))$$

Let $\theta = \lambda\varphi \in A_m \{e_1\}_A(\varphi, \varphi_1, \ldots, \varphi_k)$.
We define θ in the maximal type-structure in the following way:

If $\phi \in [\varphi]$ for some $\varphi \in A_m$, we will by the induction hypothesis have that

$$\{e_1\}_A(\varphi, \varphi_1, \ldots, \varphi_k) \simeq \{\rho(e_1)\}(\phi, \phi_1, \ldots, \phi_k, \Psi_1, \ldots, \Psi_{n-2})$$

and we let

$$\theta(\phi) = \{\rho(e_1)\}(\phi, \phi_1, \ldots, \phi_k, \Psi_1, \ldots, \Psi_{n-2}) = \theta(\varphi)$$

If there is no φ such that $\phi \in [\varphi]$, we let $\theta(\phi) = 0$. By construction we see that $\theta \in [\theta]$.
We may use Ψ_m to decide between the two cases, so uniformly from $\rho(e_1)$

we find an index e' for θ from $\Phi_1,\ldots,\Phi_k,\Psi_1,\ldots,\Psi_{n-2}$, and we let $\rho(e) = \langle 8,e',\langle\sigma'\rangle\rangle$ where σ' has the suitable form. Since $\theta \in [\theta]$ we will have that $\Phi_1(\theta) = \varphi_1(\theta)$, so

$$\{\rho(e)\}(\Phi_1,\ldots,\Phi_k,\Psi_1,\ldots,\Psi_{n-2}) = \{e\}_A(\varphi_1,\ldots,\varphi_k)$$

This ends the proof of the claim and the theorem.

□

Lemma 2.7

Let $h \in Tp(1)$ and assume

$$\theta = \lambda\Phi \in Tp(n)\{e\}(\Phi,h)$$

is total. Let $A = \langle A_n\rangle_{n\in\omega}$ be a type-structure closed under computations such that $h \in A_1$.
Then

$$\theta = \lambda\varphi \in A_n\{e\}_A(\varphi,h)$$

is total, and $\theta \in [\theta]$.

Proof: This is a direct consequence of Theorem 2.4.

□

Theorem 2.8

Let θ be computable in h . Then $2\text{-en}(\theta) \subseteq \Pi_1^1(h)$.

Proof: We will not prove this theorem in full detail. The argument is partly model-theoretic and we will indicate the main ideas. For further details, see Bergstra [1] or Moldestad-Normann [28].

Let $B \in 2\text{-en}(\theta)$, i.e. there is an index e such that

$$\alpha \in B \leftrightarrow \{e\}(\alpha,\theta) \simeq 0$$

θ is computable in h , so there is an index d such that

$$\theta = \lambda\Phi\{d\}(h,\Phi)$$

Claim

$\alpha \in B \leftrightarrow \forall A$ (if A is a countable type-structure closed under computations, $\alpha \in A_1$ and $h \in A_1$ then
$\{e\}_A(\alpha,\lambda\varphi\{d\}_A(h,\varphi)) \simeq 0$)

Proof of claim

$\rightarrow$ is a direct consequence of Theorem 2.4 and Lemma 2.7.

← Assume that α satisfies the r.h.s. above.

By the Skolem-Löwenheim theorem, let A be a countable elementary substructure of the maximal type-structure such that $\alpha, h \in A_1$.
By assumption on α

$$\{e\}_A(\alpha, \lambda\varphi\{d\}_A(h,\varphi)) \simeq 0$$

Since $A = \langle A_n \rangle_{n \in \omega}$ and $\langle Tp(n) \rangle_{n \in \omega}$ are elementary equivalent we will have that

$$\{e\}(\alpha, \lambda\phi\{d\}(h,\phi)) \simeq 0$$

But this shows that $\alpha \in B$, and the claim is proved.

In order to end the proof of this theorem, we have to transform the right hand side in the claim to a $\pi_1^1(h)$-statement. This involves a method of coding countable type-structures as elements of $Tp(1)$, so that the quantifier $\forall A$ (A is a countable type-structure ...) may be transformed to a universal function quantifier. We will not go into any details here.

□

Definition 2.9

A <u>tree</u> on ω is a nonempty set T of finite sequences σ of natural numbers such that if $\sigma \in T$ then all subsequences of σ are in T .

A tree on ω is <u>wellfounded</u> if for all $\alpha : \omega \to \omega$ there is a number n such that $(\alpha(0), ..., \alpha(n)) \in T$.

Lemma 2.10

The set of (characteristic functions of) well-founded trees is semi-computable in $^2 0$.

<u>Proof</u>: If $\sigma = (a_0, ..., a_k)$ we let $\sigma^\frown n$ be the sequence $(a_0, ..., a_k, n)$.
By the recursion theorem, let

$$\{e\}(^20,T,\sigma) = {^20}\left(\lambda n \begin{cases} \{e\}(^20,T,\sigma^\frown n) & \text{if } \sigma^\frown n \in T \\ 0 & \text{if } \sigma^\frown n \notin T \end{cases}\right)$$

If T is well-founded we prove by induction on the ordinal rank of $\sigma \in T$ that $\{e\}(^20,T,\sigma) = 0$, so in particular $\{e\}(^20,T,<>) = 0$.
On the other hand, assume that T is not well-founded and choose α such that $(\alpha(0), ..., \alpha(n)) \in T$ for all n . If for some n $\{e\}(^20,T,<\alpha(0),...,\alpha(n)>) \downarrow$ then $\{e\}(^20,T,<\alpha(0),...,\alpha(n+1)>)$ will be a subcomputation. (This is a consequence of the proof of the recursion theorem.) Thus we would have a descending sequence in the computation

tree, which is impossible. It follows that $\{e\}(^20,T,<>)\downarrow$. But then

$$T \text{ is well-founded } \longleftrightarrow \{e\}(^20,T,<>) \simeq 0$$

and the lemma is proved.

□

Corollary 2.11

a Let $^k0$ be the constant zero functional of type $k \geq 2$. Then $2\text{-en}(^k0) = \pi_1^1$.

b Let $B \subseteq Tp(1) \times Tp(k)$ be semi-computable. Define A by

$$\alpha \in A \longleftrightarrow \forall \psi \in Tp(k) \ (\alpha,\psi) \in B$$

Then $A \in \pi_1^1$.

Proof:

a From Theorem 2.8 we know that $2\text{-en}(^k0) \subseteq \pi_1^1$.

In order to prove the inclusion the other way, we must divide the argument in two cases.

Case 1 $\underline{k = 2}$: If A is π_1^1 , there is a recursive function f such that

$$\alpha \in A \longleftrightarrow f(\alpha) \text{ is a well-founded tree on } \omega .$$

So, if we let e be as in lemma 2.10

$$\alpha \in A \longleftrightarrow \{e\}(^20,f(\alpha),<>) \simeq 0$$

This shows that A is semicomputable in ^{20} so $\pi_1^1 \subseteq 2\text{-en}(^20)$.

Case 2 $\underline{k > 2}$: Let A be π_1^1 and let B be Σ_1^0 such that

$$\alpha \in A \longleftrightarrow \forall \beta \in Tp(1) \ (\alpha,\beta) \in B$$

B is semicomputable (in ordinary recursion theory) so there is an index e_0 such that $(\alpha,\beta) \in B \longleftrightarrow \{e_0\}(\alpha,\beta)\downarrow$.
Then

$$\alpha \in A \longleftrightarrow \forall \beta\{e_0\}(\alpha,\beta)\downarrow$$

$$\longleftrightarrow \forall \psi \in Tp(k-2)\{e_0\}(\alpha,P_{k-2}^1(\psi))\downarrow$$

$$\longleftrightarrow \lambda\psi\{e_0\}(\alpha,P_{k-2}^1(\psi)) \text{ is total}$$

$$\longleftrightarrow {}^k0(\lambda\psi\{e_0\}(\alpha,P_{k-2}^1(\psi))) \simeq 0$$

where P_{k-2}^1 is the push-down from definition 1.7.
This shows that A is semicomputable in $^k0$.

b Let $(\alpha,\psi) \in B \longleftrightarrow \{e\}(\alpha,\psi)\downarrow$

Then

$$\alpha \in A \longleftrightarrow \forall \psi \{e\}(\alpha,\psi)\downarrow$$

$$\longleftrightarrow \lambda\psi\{e\}(\alpha,\psi) \text{ is total}$$

$$\longleftrightarrow {}^{k+2}0(\lambda\psi\{e\}(\alpha,\psi)) \simeq 0$$

This shows that A is semicomputable in ${}^{k+2}0$, and by corollary 2.11.$\underline{a}$ we see that $A \in \Pi_1^1$.

□

2.2   The countable functionals

When we extended ordinary recursion theory to computations in functionals of higher types, we lost one of the important aspects of ordinary
recursion theory: we do not deal with just finite entities any more.
Even functionals F of type 2 act on infinite arguments. But we know
from ordinary recursive function theory that if we introduce recursive
functionals of type 2 via oracles, $F(f)$ will only depend on a finite
amount of information from f . This will be the case for the Kleene-
computable functionals of type 2 as well, but before we show this, we
need some standard notation concerning finite sequences of numbers.

Definition 2.12

$\underline{a}$ Let $f:\omega \to \omega$, $n \in \omega$. By $\bar{f}(n)$ we mean the sequence number
$<f(0),\ldots,f(n-1)>$.

$\underline{b}$ If σ is a sequence number, then $\sigma(i)$ will denote the $i+1$'th
element of the sequence. $lh(\sigma)$ will denote the length of the
sequence.
If k is a number, we let $\bar{k}(i) = k$ for all i .

Remark 2.13

$\underline{a}$ If σ is a sequence number, we see that

$$\sigma = <\sigma(0),\ldots,\sigma(lh(\sigma)-1)>$$

$\underline{b}$ $\bar{f}(n)$, $\sigma(i)$ and $lh(\sigma)$ are primitive recursive as functions of
f, n, σ and i .

Lemma 2.14

Let $f_1,\ldots,f_m \in Tp(0) \cup Tp(1)$ and let $\{e\}(f_1,\ldots,f_m) \simeq k$. Then
there is a number n such that for all $g_1,\ldots,g_m$, if $\bar{g}_i(n) = \bar{f}_i(n)$
for $i = 1,\ldots,m$, then we have that

$$\{e\}(g_1,\ldots,g_m) \simeq k$$

<u>Proof</u>: This is proved by induction on the length of the computation $\{e\}(f_1,\ldots,f_m)$. Note that S8 will not be used in such computations.

If $\{e\}(f_1,\ldots,f_m) \simeq k$ by an initial computation using S1-3, then we may put $n = 0$.

If $\{e\}(f_1,\ldots,f_m) = f_2(f_1)$ (i.e. $f_2 \in Tp(1), f_1 \in \omega$), then S6 is used. We let $n = f_1+1$.

The rest of the cases are trivial and left for the reader.

◻

<u>Corollary 2.15</u>

If $F:Tp(1) \to \omega$ is Kleene-computable in some function of type 1 , then for all $f \in Tp(1)$ there is a natural number n such that

$$\forall g(\bar{f}(n) = \bar{g}(n) \Rightarrow F(f) = F(g))$$

so $F(f)$ depends only on $\bar{f}(n)$.

◻

Corollary 2.15 actually gives a continuity-property of type 2 functionals computable in a function. In this paragraph we will make this precise.

$Tp(1) = \{f; f$ maps ω into $\omega\}$. This may be viewed as a countable product of the natural numbers. If we give ω discrete topology, we obtain the usual product topology on $Tp(1)$.

<u>Definition 2.16</u>

Let σ be a sequence-number. By B_σ we mean

$$B_\sigma = \{f:\omega \to \omega ; \bar{f}(lh(\sigma)) = \sigma\}$$
$$= \{f ; \exists n \; \bar{f}(n) = \sigma\}$$

<u>Remark 2.17</u>

It is easy to see that $\{B_\sigma\}_{\sigma \in SEC}$ is a countable basis of closed-open sets on $Tp(1)$, where SEC is the set of sequence-numbers.

<u>Lemma 2.18</u>

The following are equivalent

<u>i</u> $F:Tp(1) \to \omega$ is continuous

<u>ii</u> For all f there is an $n \in \omega$ such that

$$\forall g(\bar{g}(n) = \bar{f}(n) \Rightarrow F(g) = F(f))$$

Proof: Let F be continuous, $f \in Tp(1)$, $k = F(f)$. Then $B = F^{-1}\{k\}$ is open and $f \in B$. By remark 2.17, there is a sequence-number σ such that $f \in B_\sigma \subseteq B$. Choose n such that $\bar{f}(n) = \sigma$. For any g, if $\bar{g}(n) = \sigma$, then $g \in B_\sigma$ so $F(g) = k$. This proves i $\rightarrow$ ii.

Now, assume ii. We will show that $F^{-1}\{k\}$ is open for each k. So, let $f \in F^{-1}\{k\}$. Choose n according to ii. Then $B_{\bar{f}(n)} \subseteq F^{-1}\{k\}$, by ii. It follows that f is an element of a basis-set contained in $F^{-1}\{k\}$, so $F^{-1}\{k\}$ must be open and F must be continuous. □

Let F be continuous. If we take the sequence-numbers σ such that F is constant on B_σ and associate them with the constant value of F on B_σ, then we have given a complete description of F. Actually, if we for each f pick out some n such that F is constant on $B_{\bar{f}(n)}$ and associate with $\bar{f}(n)$ the constant value, then we have given a complete description of F. This is made precise in the following definition.

Definition 2.19

Let $F:Tp(1) \rightarrow \omega$ be given, $\alpha \in Tp(1)$.

α is called an associate for F if

i   For any sequence-numbers σ, τ, if $\sigma \prec \tau$ and $\alpha(\sigma) > 0$ then $\alpha(\tau) = \alpha(\sigma)$.

ii   If $\alpha(\sigma) > 0$ then F is constant on B_σ with value $\alpha(\sigma) - 1$.

iii $\forall f \, \exists n \, \alpha(\bar{f}(n)) > 0$

Here $\sigma \prec \tau$ means that the sequence with number τ extends the sequence with number σ.

Lemma 2.20

Let $F \in Tp(2)$. Then F is continuous if and only if F has an associate.

Proof: If F is continuous, let

$$\alpha(\sigma) = \begin{cases} 0 & \text{if } F \text{ is not constant on } B_\sigma \\ k+1 & \text{if } F \text{ is constant } k \text{ on } B_\sigma \end{cases}$$

for all sequence-numbers σ.

It is easy to see that α is an associate for F. On the other hand, if F has an associate α and $f \in Tp(1)$, choose n such that $\alpha(\bar{f}(n)) > 0$. Then F is constant on $B_{\bar{f}(n)}$, so by Lemma 2.18, F is continuous. □

Remark 2.21

Let $F:Tp(1) \to \omega$ be continuous. The associate α for F we constructed in the proof of Lemma 2.20 is called the <u>principal associate</u> for F. As for all associates for F we know that if $\alpha(\sigma) > 0$, then F is constant on B_σ. But we also know that if $\alpha(\sigma) = 0$, then F is not constant on B_σ. The principal associate is the only one with this property. We will see that there is a good reason why we defined the associates for F as we did, instead of just choosing the principal one.

Going back tó Lemma 2.14 we will see that the n needed there was actually computable from the arguments. It is easy via the argument of Lemma 2.14 to show that any functional $F:Tp(1) \to \omega$ computable in f has an associate recursive in f. The associate we obtain through this argument is, however, not always principal, and there are computable functionals such that the principal associate is not recursive. We summarize this by

Lemma 2.22

<u>a</u> $F:Tp(1) \to \omega$ is computable if and only if F has a recursive associate.

<u>b</u>   There is a computable F such that the principal associate for F is not recursive.

Proof:

<u>a</u>   F computable $\Rightarrow$ F has a recursive associate was proved in Remark 2.21.

Let α be a recursive associate for F. Then, for any f

$$F(f) = \alpha(\overline{f}(\mu n(\alpha(\overline{f}(n)) > 0))) - 1$$

where 'μn' means 'the least n such that'. Then by ordinary recursion theory F is computable in α and thus computable.

<u>b</u>   Define $F(f)$ by the following algorithm:

Let $e = f(0)$, $n = f(1)$, $<e>$ the sequence-number of the one-point sequence (e).

If $\{e\}(<e>) = 1$ by a computation in at most n steps, let $F(f) = 1$, otherwise let $F(f) = 0$. F is clearly computable. Now assume that its principal associate is recursive, say by index e_0, i.e. $\lambda\sigma\{e_0\}(\sigma)$ is the principal associate for F. If $\{e_0\}(<e_0>) = 1$, let n be the length of this computation. Choose f such that $f(0) = e_0$, $f(1) = n$. Then $f \in B_{<e_0>}$ and $F(f) = 1$. But then $\{e_0\}(<e_0>)$ should be either

0 or 2 , a contradiction.

If $\{e_0\}(<e_0>) \neq 1$, we see that F is constant zero on $B_{<e_0>}$, and the principal associate for F takes value 1 at $<e_0>$, again a contradiction.

□

The most important facts established in this paragraph are contained in the following theorem.

Theorem 2.23

Let $F:Tp(1) \to \omega$. Then the following are equivalent.

i F is continuous

ii F is computable in some $f \in Tp(1)$

iii F has an associate.

Proof:

i ⟷ iii   see Lemma 2.20

iii → ii   is a relativized version of Lemma 2.22

ii → i   see Corollary 2.15 and Lemma 2.18.

□

Recall the definition of ^2E from section 1.3. $^2E^{-1}\{0\}$ is a singleton containing the constant zero function, so $^2E^{-1}\{0\}$ is not open. Thus ^2E is not continuous. We can use ^2E to give the following recursion theoretic characterization of the continuous functionals of type 2 . The characterization is due to Grilliot [14].

Theorem 2.24

Let $F \in Tp(2)$. The following are equivalent.

i F is not continuous.

ii   There is an $f \in Tp(1)$ such that ^2E is computable in F , f .

Proof:

ii → i Assume that F is continuous. Let α be an associate for F . Then F is computable in α . If ^2E is computable in F , f for some f , then ^2E is computable in α , f . But by Theorem 2.23 ^2E would be continuous, which is impossible.

i → ii Assume that F is not continuous. Choose $f \in Tp(1)$ and a sequence $\{f_i\}_{i \in \omega}$ such that

$$f = \lim_{i \to \infty} f_i \qquad \text{while} \qquad \neg(F(f) = \lim_{i \to \infty} F(f_i))$$

This is possible since F is discontinuous.

By, if necessary, picking a subsequence we may assume

_i_ $F(f_i) \neq F(f)$ for all $i \in \omega$

_ii_ $j \geq i \to f_j(i) = f(i)$ for all $i, j \in \omega$

We will see that ^2E is computable in f, $\langle f_i \rangle_{i \in \omega}$, F.

Let $g \in Tp(1)$ be given. Define α_g by

$$\alpha_g(i) = \begin{cases} f_j(i) & \text{if } j < i \text{ is minimal such that } g(j) > 0 \\ f_i(i) & \text{if there is no } j < i \text{ such that } g(j) > 0 \end{cases}$$

It is easily seen that

$$\exists j \; g(j) > 0 \to \alpha_g = f_j \quad \text{for the least such } j$$

$$\forall j \; g(j) = 0 \to \alpha_g = f$$

Thus $^2E(g) = 0 \longleftrightarrow F(\alpha_g) = F(f)$

α_g is uniformly recursive in $g, \langle f_i \rangle_{i \in \omega}$ and it follows that ^2E is computable in $F, f, \langle f_i \rangle_{i \in \omega}$, in fact in $F, \langle f_i \rangle_{i \in \omega}$ since $f(i) = f_i(i)$ and thus f is recursive in $\langle f_i \rangle_{i \in \omega}$.   □

In the previous paragraphs we asked for a subclass of the type-two functionals that are faithful to the principle that only a finite amount of information is used in computations, and we came up with the continuous functionals of type 2. We also showed that by means of the associates we can describe these functionals using only a countable amount of information. Therefore we may also call the functionals countable. A suitable abreviation for both names is $Ct(2)$, the set of countable (continuous) functionals of type 2.

The continuous functionals will be a proper subclass of the non-normal functionals. We will later see that many properties of the continuous functionals are shared by all non-normal functionals.

Our task is now to carry this process on to higher types. If we are going to build a type-structure $\langle Ct(n) \rangle_{n \in \omega}$ preserving the idea of finiteness, any element ϕ of $Ct(3)$ must be a map $\phi : Ct(2) \to \omega$.

The first attempt could be to let $\phi \in Ct(3)$ if and only if whenever $\phi(F) = k$ then there are $f_1, \ldots, f_n$ such that for all G, if $G(f_1) = F(f_1), \ldots, G(f_n) = F(f_n)$, then $\phi(G) = \phi(F)$.

The main problem with this definition is that it does not work in the sense that there would even be computable $\phi:Ct(2) \to \omega$ not satisfying this definition:

Lemma 2.25

There is a computable function $\phi:Ct(2) \to \omega$ such that for all $F \in Ct(2)$ there is no finite set $f_1,\ldots,f_n$ such that $\phi(F)$ is determined by the values $F(f_1),\ldots,F(f_n)$.

<u>Proof:</u>

Let α_n be defined by $\alpha_n(i) = \begin{cases} 0 & \text{if } i \leq n \\ 1 & \text{if } i > n \end{cases}$

Let $\alpha(i) = 0$ for all i .

Let $\phi(F) = \mu n \, \forall m \geq n \, (F(\alpha_m) = F(\alpha))$.

Given F and $f_1,\ldots,f_k$ there will be some $m \geq \phi(F)$ such that $\alpha_m \neq f_i$ for $i = 1,\ldots,k$. Then let G be continuous such that $G(\alpha) = F(\alpha)$, $G(f_i) = F(f_i)$ for $i = 1,\ldots,k$, while $G(\alpha_m) \neq F(\alpha_m)$. Then $\phi(F) \neq \phi(G)$.

We end the proof by showing that ϕ is computable. Here we use a trick similar to the one used in the proof of Theorem 2.24. First we show that the set

$$\{(n,F);\ \exists m \geq n\ F(\alpha_m) \neq F(\alpha)\}$$

is computable.
Define

$$\nu(k) = \begin{cases} 0 & \text{if } \exists m(n \leq m \leq k \wedge F(\alpha_m) \neq F(\alpha)) \\ 1 & \text{otherwise} \end{cases}$$

Then

$$\exists m \geq n(F(\alpha_m) \neq F(\alpha)) \to \nu = \alpha_m \text{ for the least such } m$$

$$\forall m \geq n(F(\alpha_m) = F(\alpha)) \to \nu = \alpha .$$

So

$$\exists m \geq n(F(\alpha_m) \neq F(\alpha)) \longleftrightarrow F(\nu) \neq F(\alpha) .$$

ν is computable in F , so the set defined above is computable. Let $\phi(F)$ be the least n such that (n,F) is not in the set. $\square$

Since this naive approach did not work, we follow the strategy used for type 2 , we see what sort of finiteness-properties computations in continuous type 2 objects satisfy.

A typical part of a computation will be

$$\{e\}(F_1,\ldots,F_k) = F_1(\lambda x\{e_1\}(x,F_1,\ldots,F_k))$$

There may be many such applications of F_1 in a computation-tree. On the other hand, we only need to know $\lambda x \{e_1\}(x, F_1, \ldots, F_k)$ on a finite part, say for $x \in \{0, \ldots, n\}$. For any associate α_1 for F_1 we only need to know α_1 at one point. So let $\alpha_1, \ldots, \alpha_k$ be associates for $F_1, \ldots, F_k$. By a simple induction on the length of the computation $\{e\}(F_1, \ldots, F_k)$ we may show that the value is decided from a finite part of these associates, the hardest step is actually given above.

These considerations will now be turned into a formal definition of the countable functionals, $<Ct(n)>_{n \in \omega}$ and of $<As(n)>_{n \in \omega}$, the class of associates.

Definition 2.26

$Ct(0) = \omega$. Numbers are their own associates.
$Ct(1) = Tp(1)$. Functions are their own associates.

Let $\psi : Ct(n) \to \omega$ and assume that the class of associates for elements of $Ct(n)$ is defined. $\alpha : \omega \to \omega$ is an associate for ψ if and only if for each $\varphi \in Ct(n)$ and for each associate β for φ there is a minimal k such that $\alpha(\bar{\beta}(k)) > 0$, and for all $i \geq$ this k

$$\alpha(\bar{\beta}(i)) = \psi(\varphi) + 1$$

$\psi \in Ct(n+1)$ if and only if ψ has an associate.

Remark 2.27

<u>a</u>   We call the functionals in $<Ct(n)>_{n \in \omega}$ <u>countable</u> since they are described by countably much information with the aid of the associates.

<u>b</u>   The countable functionals were independently introduced by Kleene [23] and Kreisel [24] as a good basis for recursion theory and as a valuable tool in intuitionistic logic and constructive analysis. Kleene defined his associates as we did, but his countable functionals are elements in the maximal type-structure. Kreisel did not presuppose the existence of higher type functionals, and his <u>continuous</u> functionals are equivalence classes of associates. Our version is isomorphic to Kreisels model for the continuous functionals.

<u>c</u>   In chapter 3 we will analyze the topological aspects of $<Ct(n)>_{n \in \omega}$ and justify the term <u>continuous</u> functional.

<u>d</u>   The associates may be viewed as a sort of abstract algorithms for the functionals, algorithms not working on functionals of lower types but on algorithms for functionals of lower types. This view will be investigated further in the next section.

Our reason for not accepting the first definition of Ct(3) was that it would not be closed under computations. We are now going to show that $\langle Ct(n)\rangle_{n\in\omega}$ as we have defined it is closed under computations. This result is due to Kleene [23].

Theorem 2.28

There is a partial recursive function $f:\omega \to \omega$ such that whenever $\alpha_1,\ldots,\alpha_n$ are associates for the countable functionals $\varphi_1,\ldots,\varphi_n$ and

$$\{e\}(\varphi_1,\ldots,\varphi_n) \simeq m$$

then there is a minimal k_0 depending on $e,\alpha_1,\ldots,\alpha_n$ such that

$$f(\langle e,\bar{\alpha}_1(k_0),\ldots,\bar{\alpha}_n(k_0)\rangle) > 0$$

and for all $k \geq k_0$

$$f(\langle e,\bar{\alpha}_1(k),\ldots,\bar{\alpha}_n(k)\rangle) = m+1$$

Moreover

$$f(\langle e,\bar{\alpha}_1(k),\ldots,\bar{\alpha}_n(k)\rangle)$$

will be defined for all k.

Proof: In the proof, recall the convention that if φ_i is a number s, then $\alpha_i = s$ and $\bar{\alpha}_i(k) = s$ for all k.

We will define f by the recursion theorem. The proof will then be by induction on the length of the computation $\{e\}(\varphi_1,\ldots,\varphi_n)$. As usual we define f and prove that it works simultaneously. As an induction hypothesis we assume that f works for all subcomputations of the one we consider.

S1 $\{e\}(x,\varphi_1,\ldots,\varphi_n) = x+1$
     Define $f(\langle e,x,\bar{\alpha}_1(k),\ldots,\bar{\alpha}_n(k)\rangle) = x+2$ for all α,k.
     This clearly works.

S2 $\{e\}(\varphi_1,\ldots,\varphi_n) = q$
     Let $f(\langle e,\bar{\alpha}_1(k),\ldots,\bar{\alpha}_n(k)\rangle) = q+1$ for all α,k.

S3 $\{e\}(x,\varphi_1,\ldots,\varphi_n) = x$
     Let $f(\langle e,x,\bar{\alpha}_1(k),\ldots,\bar{\alpha}_n(k)\rangle) = x+1$ for all x,α,k.

S4 $\{e\}(\varphi_1,\ldots,\varphi_n) \simeq \{e_1\}(\{e_2\}(\varphi_1,\ldots,\varphi_n),\varphi_1,\ldots,\varphi_n)$
     Let

$$f(\langle e,\bar{\alpha}_1(k),\ldots,\bar{\alpha}_n(k)\rangle) = \begin{cases} f(\langle e_1,f(\langle e_2,\bar{\alpha}_1(k),\ldots,\bar{\alpha}_n(k)\rangle)-1,\bar{\alpha}_1(k),\ldots,\bar{\alpha}_n(k)\rangle) \\ \quad \text{if } f(\langle e_2,\bar{\alpha}_1(k),\ldots,\bar{\alpha}_n(k)\rangle) > 0 \\ 0 \quad \text{otherwise} \end{cases}$$

By the induction hypothesis there is a minimal k_0 such that for
$x = \{e_2\}(\varphi_1,\ldots,\varphi_n)$, $y = \{e_1\}(x,\varphi_1,\ldots,\varphi_n)$ we have

$$f(<e_2,\bar{a}_1(k_0),\ldots,\bar{a}_n(k_0)>) = x+1 \quad \& \quad f(<e_1,x,\bar{a}_1(k_0),\ldots,\bar{a}_n(k_0)>) = y+1$$

For $k < k_0$ we see that $f(<e,\bar{a}_1(k),\ldots,\bar{a}_n(k)>) = 0$ while for $k \geq k_0$
it follows from the definition and the induction hypothesis that

$$f(<e,\bar{a}_1(k),\ldots,\bar{a}_n(k)>) = \{e\}(\varphi_1,\ldots,\varphi_n) + 1$$

<u>S5</u> $\{e\}(0,\varphi_1,\ldots,\varphi_n) = \{e_1\}(\varphi_1,\ldots,\varphi_n)$

$\{e\}(x+1,\varphi_1,\ldots,\varphi_n) = \{e_2\}(\{e_1\}(x,\varphi_1,\ldots,\varphi_n),x,\varphi_1,\ldots,\varphi_n)$

Define $f(<e,x,\bar{a}_1(k),\ldots,\bar{a}_n(k)>)$ by

$$\begin{cases} f(<e_2,\bar{a}_1(k),\ldots,\bar{a}_n(k)>) & \text{if } x = 0 \\ f(<e_2,f(<e,x-1,\bar{a}_1(k),\ldots,\bar{a}_n(k)>)-1,x-1,\bar{a}_1(k),\ldots,\bar{a}_n(k)>) \\ \quad \text{if } x > 0 \text{ and } f(<e,x-1,\bar{a}_1(k),\ldots,\bar{a}_n(k)>) > 0 \\ 0 \quad \text{otherwise} \end{cases}$$

The proof that this works follows the same pattern as <u>S4</u> .

<u>S6</u> $\{e\}(\varphi_1,\ldots,\varphi_n) = \{e_1\}(\varphi_{\tau(1)},\ldots,\varphi_{\tau(n)})$

Let $f(<e,\bar{a}_1(k),\ldots,\bar{a}_n(k)>) = f(<e_1,\bar{a}_{\tau(1)}(k),\ldots,\bar{a}_{\tau(n)}(k)>)$

This clearly works.

<u>S7</u> $\{e\}(x,\varphi_1,\ldots,\varphi_n) = \varphi_1(x)$, where $\varphi_1 \in \mathrm{Tp}(1)$.

Let

$$f(<e,x,\bar{a}_1(k),\ldots,\bar{a}_n(k)>) = \begin{cases} \alpha_1(x)+1 & \text{if } x < k \\ 0 & \text{if } x \geq k \end{cases}$$

Here we see that $k_0 = x+1$.

<u>S8</u> $\{e\}(\varphi_1,\ldots,\varphi_n) = \varphi_1(\lambda\varphi\{e_1\}(\varphi,\varphi_1,\ldots,\varphi_n))$

Define

$$\beta(\pi) = f(<e_1,\pi,\bar{a}_1(\mathrm{lh}(\pi)),\ldots,\bar{a}_n(\mathrm{lh}(\pi))>)$$

<u>Claim</u> β is an associate for $\lambda\varphi\{e_1\}(\varphi,\varphi_1,\ldots,\varphi_n)$.

<u>Proof of claim</u>: Let α be an associate for φ . By the induction-
-hypothesis there is a minimal k_0 such that

$$f(<e_1,\bar{a}(k_0),\bar{a}_1(k_0),\ldots,\bar{a}_n(k_0)>) > 0$$

and for $k \geq k_0$

$$f(<e_1,\bar{a}(k),\bar{a}_1(k),\ldots,\bar{a}_n(k)>) = \{e_1\}(\varphi,\varphi_1,\ldots,\varphi_n) + 1 .$$

Then $k < k_0 \Rightarrow \beta(\bar{a}(k)) = 0$ while $k \geq k_0 \Rightarrow \beta(\bar{a}(k)) = \{e_1\}(\varphi,\varphi_1,\ldots,\varphi_n) + 1$

□ Claim

We now define $f(<e,\bar{a}_1(k),\ldots,\bar{a}_n(k)>)$ by

$$\begin{cases} m+1 & \text{if } (\exists k_1 < k)(\bar{\beta}(k_1) < k \wedge \alpha_1(\bar{\beta}(k_1)) = m+1 \\ 0 & \text{otherwise} \end{cases}$$

Since α_1 is an associate for φ_1, there will be a minimal k_1 such that $\alpha_1(\bar{\beta}(k_1)) > 0$ and by the claim, for $k \geq k_1$ we will have

$$\alpha_1(\bar{\beta}(k_1)) = \{e\}(\varphi_1,\ldots,\varphi_n) + 1$$

If we let $k_0 = \bar{\beta}(k_1) + 1$ we see that

$k < k_0 \Rightarrow f(<e,\bar{a}_1(k),\ldots,\bar{a}_n(k)>) = 0$ while

$k \geq k_0 \Rightarrow f(<e,\bar{a}_1(k),\ldots,\bar{a}_n(k)>) = \alpha_1(\bar{\beta}(k_1)) = \{e\}(\varphi_1,\ldots,\varphi_n) + 1$

<u>S9</u> $\{e\}(e_1,\varphi_1,\ldots,\varphi_n) = \{e_1\}(\varphi_1,\ldots,\varphi_t)$ $t \leq n$.

Define

$$f(<e,e_1,\bar{a}_1(k),\ldots,\bar{a}_n(k)>) = f(<e_1,\bar{a}_1(k),\ldots,\bar{a}_t(k)>)$$

This clearly works.

This ends the proof of Theorem 2.28.

□

<u>Corollary 2.29</u>

<u>a</u>   If $\psi = \lambda\varphi \in Ct(k)\{e\}(\varphi,\varphi_1,\ldots,\varphi_n)$ is total, where $\varphi_1,\ldots,\varphi_n$ are countable, then ψ is countable.

<u>b</u>   The countable functions are closed under computations.

<u>Proof</u>:

<u>a</u>   is actually what we proved in the claim under case S8.

<u>b</u>   follows trivially from <u>a</u>.

□

2.3 Countable recursion and the associates

When we defined the countable functionals, we had two guidelines. We wanted each functional ψ to utilize only a finite amount of information about an argument φ in order to decide the value $\psi(\varphi)$. Moreover, we wanted the countable functionals to be closed under computations.

We made our choice on the basis of one failure (Lemma 2.25) and the fact that we, via the associates, preserve a certain finiteness.

We could ask other questions, seeking confirmation or refutation of the naturalness of this type-structure, i.e.

i   Are there smaller type-structures containing Ct(2) that are still closed under computations ?

ii Are there larger type-structures still with some finiteness--property ?

These questions will be answered later. We will construct an element of Ct(3) which is not computable in any type-one function (Theorem 4.40). On the other hand, if we try to extend $<Ct(n)>_{n \in \omega}$ at any type, ^2E will be computable in any added functional and some countable object. (Theorem 3.31).

In generalized recursion theory the notion of finiteness is also generalized. A set A is finite in a functional ψ if and only if quantification over A is computable in ψ. So, the countable functionals is a maximal type-structure such that real finiteness and ψ-finiteness coincide for all functionals ψ.

Clearly, our solution to ii above serves as a justification for defining the countable functionals the way we did, but we must admit that these functionals existed for almost twenty years without this result. One of the original motivations was due to G. Kreisel. Intuitionistically, such functionals have no existence except through their descriptions. If we want to construct a set of functionals not accepting the power set and if we think that a functional is just the fulfillment of some intention, then the intention for one functional has nothing to work on but some intention for the argument. Making this precise, one is led to the associates, or actually to finite bits of associates acting on one another.

Another question we ought to ask is the following: Is Kleene-computability the best notion, or the only notion of computability on the countable functionals? Whether it is the best notion, we will leave to the individual taste, but there is another notion, countable recursion, which we will define. We will also discuss the difference between this new notion and Kleene-computability.

Before giving the formal definition of countable recursion, we will try to motivate it by returning to our example of computability of the Riemann-integral.

If you ask a computer to compute $\int_0^1 f(x)dx$, it will demand an algorithm for f , and if you require the computer to produce the answer

with an error less than an arbitrarily given $\varepsilon > 0$, the computer may
demand the same of your algorithm for f .

Forgetting about tricky programmers wanting to compute quickly, we
imagine that the computer wants the algorithm for f to have the following properties:

For a fixed input-number x with n decimals, the output will be
a finite decimal number y and an uncertainty-number m , and the information should be: For any number x' with n first decimals x ,

$$f(x') \in [y - \frac{1}{m} , y + \frac{1}{m}]$$

Moreover, for any number x' the uncertainty-number tends monotonely to
infinity as the number of decimals in the bit you put into the algorithm
tends to infinity.

So the computer is not interested in f , but in an 'associate' for
f estimating $f(x)$ from estimates of x .

Now, uniformly recursive in any such 'associate' for f , we may,
through upper and lower step-functions, estimate $\int_0^1 f(x)dx$ as good as
we want. Moreover, in a given estimate we only consult the 'associate'
for f finitely often.

Our example in the previous paragraph illustrates that if we ask a
machine to compute a higher type functional, it will require an algorithm
for the argument in order to give an answer. So, if we want to give a
model for real-life computations in higher types, Kleene-computation is
not the right concept, we are led to the following definition.

Definition 2.30

<u>a</u>   A functional $\psi \in Ct(k)$ is said to be <u>recursive</u> or <u>recursively
countable</u> if it has a recursive associate.

<u>b</u> ψ is <u>recursive in φ</u> or <u>recursively countable in φ</u> if there is a
recursive function $\gamma : Tp(1) \to Tp(1)$ such that whenever α is an
associate for φ then $\gamma(\alpha)$ is an associate for ψ .

<u>c</u>   The set of <u>recursions</u> or countable computations $[e](\psi_1, \ldots, \psi_n)$ is
defined as follows

$[e](\psi_1, \ldots, \psi_n) \simeq k$ if and only if for all associates

$\alpha_1, \ldots, \alpha_n$ for $\psi_1, \ldots, \psi_n$ we have that $\{e\}(\alpha_1, \ldots, \alpha_n) = k$.

Remark 2.31

<u>a</u>   From Theorem 2.28 it follows immediately that all Kleene-computable

functionals are recursive, that 'computable in' is a finer partial
ordering than 'recursive in', and that there is a recursive function ρ
such that whenever $\{e\}(\psi_1,\dots,\psi_n) \simeq k$ then $[\rho(e)](\psi_1,\dots,\psi_n) \simeq k$.

<u>b</u>   A functional F of type 2 is computable if and only if it is re-
cursive. This can not be generalized to higher types, and it cannot be relativized to functionals of type ≥ 3 .

<u>c</u>   We will use 'recursive' and 'recursion' in this book. The terms
'countable recursive' etc. have been used elsewhere in the literature.
The present terminology was suggested in Gandy-Hyland [13].

  In Remark 2.31 <u>a</u> we see that the computations may be imbedded in
the set of recursions. We will now show that the set of recursions has
some of the important properties of a general computation theory (see
Fenstad [11]). In a sense, they are 'closed under S1-S9', we will establish this fact to some extent below.

Lemma 2.32

<u>a</u>   There is a recursive function $\rho_1(e_1,e_2)$ such that

$$[\rho_1(e_1,e_2)](\psi_1,\dots,\psi_n) \simeq [e_1]([e_2](\psi_1,\dots,\psi_n),\psi_1,\dots,\psi_n)$$

<u>b</u>   There is a recursive function $\rho_2(e)$ such that whenever

$$\Phi = \lambda\varphi[e](\varphi,\psi_1,\dots,\psi_n)$$

is total, then

$$[\rho_2(e)](\psi_1,\dots,\psi_n) = \psi_1(\Phi)$$

provided the types fit.

<u>c</u>   For each n, t there is an index $e_{n,t}$ uniformly recursive in
n, t   such that for all $e',\psi_1,\dots,\psi_n$

$$[e_{n,t}](e',\psi_1,\dots,\psi_n) \simeq [e'](\psi_1,\dots,\psi_t)$$

Proof:

<u>a</u>   Let $\rho_1(e_1,e_2) = <u,e_1,e_2,<\sigma>>$ where σ gives the types for a
list of associates for $\psi_1,\dots,\psi_n$.

<u>b</u>   Construct f recursively in associates $\beta_1,\dots,\beta_n$ for $\psi_1,\dots,\psi_n$
as follows:

If $\Phi \in Tp(1)$, let $f = \Phi$. Otherwise let

$$f(\tau) = \begin{cases} 0 & \text{if } \{e\}_{lh(\tau)}(\tau,\beta_1,\dots,\beta_n) \text{ does not give a value} \\ k+1 & \text{if } \{e\}_{lh(\tau)}(\tau,\beta_1,\dots,\beta_n) \text{ gives value } k \end{cases}$$

(In $\{e\}_{lh(\tau)}(\tau,\beta_1,\ldots,\beta_n)$ we compute $lh(\tau)$ steps and treat τ as the beginning of a function.)

By the definition of Φ , f will be an associate for Φ .

Let $\rho_2(e)$ be such that

$$\{\rho_2(e)\}(\beta_1,\ldots,\beta_n) = \beta_1(\bar{f}(k)) - 1 \quad \text{where} \quad k \quad \text{is minimal}$$
$$\text{such that} \quad \beta_1(\bar{f}(k)) > 0 .$$

Since β_1 is an associate for ψ_1 and f is an associate for Φ , we see that $\{\rho_2(e)\}(\beta_1,\ldots,\beta_n) \simeq \psi_1(\Phi)$. But then $[\rho_2(e)](\psi_1,\ldots,\psi_n) = \psi_1(\Phi)$.

c Let $e_{n,t} = <9,n,t,<\sigma>>$ where σ is the list of the types of the associates for $\psi_1,\ldots,\psi_n$.

        □

Remark 2.33

In this theory of recursions, we lack a clear notion of a subcomputation and a notion of a length of a recursion. This means that we do not deal with a computation theory in the sense of Fenstad [11].

Feferman [10] and Hyland [21] asked if the theory is inductive, which roughly might mean

can the relation $[e](\psi_1,\ldots,\psi_n) \simeq k$ be defined as the closure of some inductive definition over $<Ct(n)>_{n\in\omega}$ using a finite list of generators ?

After this manuscript was typed, the author has given a positive answer to this problem.

We will now take a look at the complexity of the associates. Later, in chapter 5, we will give a more thorough analysis of the associates, after which we may give an easier proof of the following lemma.

Lemma 2.34

a The set $As(n)$ is Π^1_{n-1} .

b The relation $\alpha \simeq_n \beta$, i.e. '$\alpha \in As(n) \wedge \beta \in As(n)$ and α and β are associates for the same functional'is Π^1_{n-1} .

Proof: We prove a and b simultaneously by induction on n .

n = 1 $\alpha \in As(1) \longleftrightarrow \alpha \in Tp(1)$

   $\alpha \simeq_1 \beta \longleftrightarrow \forall n(\alpha(n) = \beta(n))$, which is Π^0_1 . If we let Π^1_0 denote the arithmetic relations, we see that the lemma is valid for n=1.

<u>n = k+1</u>

<u>a</u> $\alpha \in As(n) \longleftrightarrow \forall\beta(\beta \in As(k) \rightarrow \exists n\ \alpha(\bar{\beta}(n)) = 0 \ \wedge$

$\wedge\ \forall n,m\ (\alpha(\bar{\beta}(n)) > 0 \wedge m > n \rightarrow \alpha(\bar{\beta}(m)) = \alpha(\bar{\beta}(n))))$

$\wedge\ \forall\beta,\gamma(\beta \approx_k \gamma \rightarrow \forall n,m\ (\alpha(\bar{\beta}(m)) > 0 \wedge \alpha(\bar{\gamma}(n)) > 0$

$\rightarrow \alpha(\bar{\beta}(m)) = \alpha(\bar{\gamma}(n))))$

<u>b</u> $\alpha \approx_n \beta \longleftrightarrow \forall\gamma,n,m\ (\gamma \in As(k) \wedge \alpha(\bar{\gamma}(n)) > 0 \wedge \beta(\bar{\gamma}(m)) > 0$

$\rightarrow \alpha(\bar{\gamma}(n)) = \beta(\bar{\gamma}(m)))$

$\wedge\ \alpha \in As(n) \wedge \beta \in As(n)$

By standard manipulations of quantifiers and a suitable induction hypothesis we see that these definitions are of the appropriate form.

□

<u>Remark 2.35</u>

<u>a</u>   In fact, if α and β are elements of As(n) , then $\alpha \approx_n \beta$ is a
Π_1^0-statement. This will be discussed in chapter 5.

<u>b</u>   In chapter 5 we will show that As(n) is a complete Π_{n-1}^1-relation.
The origin of this result is lost in history, an alternative proof
is given in Hyland [21]. We will use it to characterize semirecursion.
The characterization (Theorem 2.37) is due to Hyland [21].

<u>Lemma 2.36</u>

There is a recursive map $F: Tp(1) \rightarrow Tp(1) \times Tp(1)$ that maps $As(^k 0)$
onto $As(^k 0) \times As(^k 0)$

where $^k 0$ is the functional in Ct(k) that is constant zero
and $As(^k 0)$ is its set of associates.

<u>Proof</u>: Let τ be a finite sequence. Let $<k,\tau>$ be the sequence of
the same length as τ with i'th coordinate defined as follows

$$<k,\tau>(i) = \begin{cases} 0 & \text{if } \tau(i) = 0 \\ <k,\tau(i)-1>+1 & \text{if } \tau(i) > 0 \end{cases}$$

For $\alpha \in Tp(1)$, define $<k,\alpha>$ in the same way.
Let $\gamma \in Tp(1)$. Let $\gamma_0(\tau) = \gamma(<0,\tau>)$, $\gamma_1(\tau) = \gamma(<1,\tau>)$. (Here we
regard γ as acting on finite sequences, though it is really acting on
sequence numbers.) Let $F(\gamma) = (\gamma_0,\gamma_1)$. Clearly F is recursive.
Moreover, if α is an associate for $\varphi \in Ct(k-1)$, then $<n,\alpha>$ is an
associate for $<^{k-1}n,\varphi>$, where ^{k-1}n is the constant n in Ct(k-1).

Assume that γ is an associate for k_0. We will show that γ_0 (and γ_1) is an associate for k_0. Let $\alpha \in As(k-1)$. Then $<0,\alpha> \in As(k-1)$, and for some n we have that $\gamma(\overline{<0,\alpha>}(n)) = 1$. By definition of γ_0 then $\gamma_0(\bar{\alpha}(n)) = 1$. This shows that $\gamma_0 \in As(k_0)$, and by the same argument we show that $\gamma_1 \in As(k_0)$.

We have shown $F:As(k_0) \to As(k_0) \times As(k_0)$. It remains to show that the map is onto.

Let δ_0, δ_1 be two associates for k_0. Define γ by

$$\gamma(\tau) = \begin{cases} 0 & \text{if } \tau \text{ is on the form } <0,\tau'> \text{ and } \delta_0(\tau') = 0 \\ 0 & \text{if } \tau \text{ is on the form } <1,\tau'> \text{ and } \delta_1(\tau') = 0 \\ 1 & \text{otherwise} \end{cases}$$

We see that $\gamma_0 = \delta_0$ and $\gamma_1 = \delta_1$, so $F(\gamma) = (\delta_0, \delta_1)$. This ends the proof of the lemma.

□

Theorem 2.37

Let k_0 be as in Lemma 2.36, $k \geq 0$. Let $A \subseteq Tp(1)$. Then the following are equivalent

i $A \in \pi_k^1$

ii   For some e, $\forall \alpha(\alpha \in A \leftrightarrow [e](\alpha, k_0) \simeq 0)$.

Proof: As mentioned in Remark 2.35 b we will assume that the set $As(k_0)$ is complete π_{k-1}^1. This will be shown in section 5.3 (Lemma 5.24).

ii → i : Assume that

$$\alpha \in A \leftrightarrow [e](\alpha, k_0) \simeq 0$$

Then

$$\alpha \in A \leftrightarrow \forall \beta \in As(k_0)([e](\alpha, \beta) \simeq 0)$$

This is clearly π_k^1.

i → ii : Let A be π_k^1. We will produce an e such that

$$\alpha \in A \leftrightarrow [e](\alpha, k_0, k_0) \simeq 0$$

By Lemma 2.36 this will be sufficient.
For each $\gamma: \omega \to \omega$, let

$$h_\gamma(n) = \mu k(\gamma(^1\overline{(n+1)}(k+1))) > 0 .$$

h_γ is uniformly partial recursive in γ.

Claim: When γ varies over $As(^k0)$ then h_γ varies over $Tp(1)$.

Proof: We assume $k > 2$. For $k = 2$ we may use a similar argument which we leave for the reader.

$^1(n+1)$ is an associate for the constant n-functional of type $k-1$, so if $\gamma \in As(^k0)$ there will be a k_0 such that $\gamma(^1\overline{(n+1)}(k_0)) = 1$, and $h_\gamma(n)$ is defined. So h_γ is total.

Now, let h be given. We will construct γ such that $h_\gamma = h$. Let τ be a sequence-number of positive length. If $\tau(i) = n+1$ for all $i < lh(\tau)$, then let

$$\gamma(\tau) = \begin{cases} 0 & \text{if } lh(\tau) < h(n)+1 \\ 1 & \text{if } lh(\tau) \geq h(n)+1 \end{cases}$$

otherwise let $\gamma(\tau) = 1$. Clearly $\gamma(\tau)$ is an associate for $^k0$ and $h_\gamma = h$.

This ends the proof of the claim and we return to the proof of $\underline{i \rightarrow ii}$.

Let $B \in \Sigma^1_{k-1}$ be such that

$$\alpha \in A \longleftrightarrow \forall\beta \, <\alpha,\beta> \in B$$

Now $As(^k0)$ is a complete Π^1_{k-1}-set, so there is a recursive function $F: Tp(1) \rightarrow Tp(1)$ such that

$$<\alpha,\beta> \in B \longleftrightarrow F(<\alpha,\beta>) \in As(^k0)$$

So we get

$$\alpha \in A \longleftrightarrow \forall\beta \, F(<\alpha,\beta>) \in As(^k0) \longleftrightarrow \forall\gamma \in As(^k0) \, \forall\delta \in As(^k0) \, F(<\alpha,h_\gamma>) \neq \delta$$

Let

$$\{e\}(\alpha,\gamma,\delta) = 0 \cdot \mu n[F(<\alpha,h_\gamma>)(n) \neq \delta(n)]$$

Clearly $\alpha \in A$ if and only if for all $\gamma,\delta \in As(^k0)$ we have that $\{e\}(\alpha,\gamma,\delta) = 0$, so

$$\alpha \in A \longleftrightarrow [e](\alpha,^k0,^k0) \simeq 0$$

□

Remark 2.38

This result, compared with Theorem 5.46 shows a difference between the set of computations and the set of recursions. We will show later that the complexity of the computations lies two levels lower in the projective hierarchy.

We have now some familiarity with the notion of countable recursion and the main difference between recursions and computations is clear, in computations we use algorithms that work directly on the functionals while recursions work on descriptions of the functionals, descriptions which are not themselves always computable in the functional. This is why we think both notions are worth a study, in computations we investigate the computing power of the functional itself while in recursions we investigate the uniform computing power of a description of the functional, via the associates.

When we go into a closer analysis of the associates we will see that finite bits of the associates may be regarded as finite type-structures approximating the countable functionals. So in recursions we compute on arbitrary finite approximations to the functionals, in Kleene-computations we deal with the functionals themselves.

We must admit that later, when we deal with recursion it is mainly to compare it with computability. The main reason is that the theory for recursions is not yet well developed while the investigation of computations and computability has been more successful. There is no reason to prefer the one from the other.

In comparison with recursion in normal functionals, our computations will be very effective and finitistic. Another area where effective operations are used is effective algebra.

You may either investigate recursively represented algebraic systems or you may investigate computations based on the algorithmic power of the algebraic structure itself. The latter approach was taken by H. Friedman [12] and has been investigated with success in Moldestad, Stoltenberg-Hansen, Tucker [29] and [30] and in Tucker [44].

There are some analogies with recursions and computations, in the first case you work, both in algebra and higher type recursion theory, with descriptions of the structure or object you investigate, in the second case you regard the algorithmic power expressible from the structure or functional itself. The latter approach is more delicate than the first, and thus gives the opportunity of a finer analysis of the notion of computability.

3. Ct(n) AS A TOPOLOGICAL SPACE

3.1 The topology

When we defined the countable functionals, we started off with the intention of defining a typestructure where each application $\psi(\varphi)$ only depends on a finite amount of information about φ. After some considerations, we ended up with the associates $As(n)$ and the corresponding functionals $Ct(n)$.

Clearly the precise set of associates will depend much on the basic tools we use, for instance which pairing function $<,>$ we use on ω, and how we enumerate the finite sequences. At first sight one could even fear that the classes $Ct(n)$ themselves would depend on the choice of such basic tools, a phenomenon which would be rather unpleasant.

After some thinking however, we may convince ourselves that with some combinatorial endurance we are able to formulate and prove that the classes $Ct(n)$ are invariant under change of such tools.

Still we may wish to look for other ways of defining these functionals, both to justify why we are interested in them and to investigate them further. This is what this chapter is about, how to put some structure on $Ct(n)$ and various ways of characterizing the spaces.

Much of the pioneering work here was done by Martin Hyland in [20] where he regarded $Ct(n)$ as a filterspace and as a limit space. Also many of the topological results were announced without proof in Hyland [20]. Contributions to this area, not covered by this book, are given in Ershov [7], [8] and [9]. See Hyland [20] for further references.

Our recursion-theoretic characterization of $Ct(n)$ is originally due to Bergstra [3], with a somewhat different proof.

Our first aim is to construct a topology T_k on $Ct(k)$ such that $Ct(k+1)$ really are the continuous functionals.

We recall that the sets $B_\tau = \{f; \exists n\, \bar{f}(n)=\tau\}$ form a basis of closed-open sets for the topology on $Tp(1)$. As a first attempt we lift this definition.

Definition 3.1

Let σ be a sequence-number.
Let $B_\sigma^k = \{\psi \in Ct(k); \exists\, \text{associate } \alpha \text{ for } \psi\ \exists n\,(\bar{\alpha}(n) = \sigma)\}$.

Lemma 3.2

Let $\Phi \in Ct(k+1)$, $\psi \in Ct(k)$. Then there is a sequence σ such that $\psi \in B_\sigma^k$ and Φ is constant on B_σ^k.

Proof: Let α be an associate for Φ, β an associate for ψ. Choose n such that $\alpha(\bar{\beta}(n)) > 0$ and let $\sigma = \bar{\beta}(n)$. Then by the definition of B_σ^k, $\psi \in B_\sigma^k$. Moreover, for any $\psi' \in Ct(k)$, if ψ' has an associate β' such that $\bar{\beta}'(n) = \sigma$, it follows from the definition of an associate that $\Phi(\psi') = \Phi(\psi)$. So Φ is constant $\Phi(\psi)$ on B_σ^k. □

Remark 3.3

$B_\sigma^1 = B_\sigma$ for all σ.

For $k > 1$ it is not true that given σ and τ we have

$$B_\sigma^k \subseteq B_\tau^k \ \vee \ B_\tau^k \subseteq B_\sigma^k \ \vee \ B_\tau^k \cap B_\sigma^k = \emptyset$$

In chapter 5 we will show that the following relations are primitive recursive.

Definition 3.4

Define the predicate Con (for consistent) by

$\underline{Con(k,\sigma,\tau)}$ if $B_\sigma^k \cap B_\tau^k \neq \emptyset$

$\underline{Con\ (k,\sigma)}$ if $Con(k,\sigma,\sigma)$

Remark 3.5

$Con(k,\sigma,\tau)$ means that there is a functional ψ of type k with associates α and β that extend σ and τ respectively.

Our next result shows that when $k > 1$ we cannot use the B_σ^k's as a basis for a suitable topology on $Ct(k)$. We will give the complete proof for $k = 2$ and just indicate how it goes for $k > 2$.

Lemma 3.6

Let $k > 1$. There is a functional $\Phi : Ct(k) \to \omega$ such that

$\underline{i}$ Φ has no associate.

$\underline{ii}$ For each $\psi \in Ct(k)$ there is a σ such that $\psi \in B_\sigma^k$ and Φ is constant on B_σ^k.

Proof: Let

$$\Phi(\psi) = \begin{cases} 0 & \text{if } \psi \text{ is constant } 0 \\ 1 & \text{otherwise} \end{cases}$$

Proof of i : Assume that α is an associate for Φ. Let β be an associate for the constant 0 functional ψ such that

If $k = 2$ then $\beta(< >) = 0$

If $k > 2$ then $\beta(\tau) > 0 \land \text{Con}(k-1,\tau) \Rightarrow \exists\sigma(\text{Con}(k-2,\sigma) \land \sigma < \text{lh}(\tau) \land \tau(\sigma) > 0)$

We split the argument in two cases.

<u>$k = 2$</u>   Let n be such that $\alpha(\bar{\beta}(n)) = 1$. $\bar{\beta}(n)$ will only contain information about ψ from some finite union $B_{<0>} \cup \ldots \cup B_{<m>}$. So let ψ' be zero on $B_{<0>} \cup \ldots \cup B_{<m>}$ and one elsewhere, and let β' be an associate for ψ' such that $\bar{\beta}'(n) = \bar{\beta}(n)$. But then $\alpha(\bar{\beta}'(n)) \in \{0,2\}$, a contradiction.

<u>$k > 2$</u>   Again, choose n such that $\alpha(\bar{\beta}(n)) = 1$. We must construct an associate β' for a functional $\psi' \neq \psi$ such that $\bar{\beta}'(n) = \bar{\beta}(n)$.
Let

$$t = \max\{\tau(\sigma); \tau < n \land \beta(\tau) > 0 \land \text{Con}(k-1,\tau) \land \text{Con}(k-2,\sigma)$$
$$\land \sigma < \text{lh}(\tau) \land \tau(\sigma) > 0\}$$

Let φ be the constant 0 functional of type $k-2$. Let

$$\psi'(\Psi) = \begin{cases} 0 & \text{if } \Psi(\varphi) \leq t \\ 1 & \text{if } \Psi(\varphi) > t \end{cases}$$

The idea is that ψ' and ψ coincide on the part of ψ described by $\bar{\beta}(n)$. It is tedious but not difficult to write down a proper definition of β' .

<u>Proof of ii</u> :   If ψ is not constant zero and β is an associate for ψ choose a τ such that $\text{Con}(k-1,\tau)$ and $\beta(\tau) > 1$. Choose $n > \tau$. Then $\psi \in B^k_{\bar{\beta}(n)}$ and ϕ is constant 1 on $B^k_{\bar{\beta}(n)}$.

If ψ is constant zero, let β be the constant 1 function. Then β is an associate for ψ , and since $\beta(< >) = 1$, we see that for any n larger than the code of the empty sequence we have

$$B^k_{\bar{\beta}(n)} = \{\psi\} , \text{ so } \psi \in B^k_{\bar{\beta}(n)} \text{ and } \phi \text{ is constant zero on } B^k_{\bar{\beta}(n)} .$$

□

In the proof of Lemma 3.2 we could use any associate β for ψ , so the proof gave more information than the actual lemma. Let us use the full strength of the lemma now.

<u>Lemma 3.7</u>
Let $\phi: \text{Ct}(k) \to \omega$. Then the following are equivalent.

<u>i</u> $\phi \in \text{Ct}(k+1)$
<u>ii</u> $\forall\psi \in \text{Ct}(k) \, \forall$ associates β for $\psi \, \exists n \, (\phi$ is constant on $B^k_{\bar{\beta}(n)})$.

Proof: <u>i→ii</u> was proved in lemma 3.2. So assume <u>ii</u>.
Then

$$\alpha(\tau) = \begin{cases} t+1 & \text{if } \varphi \text{ is constant } t \text{ on } B_\tau^k \\ 0 & \text{otherwise} \end{cases}$$

is easily seen to be an associate for φ.

This lemma suggests that a set $A \subseteq Ct(n)$ should be open if when-
ever $\psi \in A$ and β is an associate for ψ then for some n , $B_{\beta(n)}^k \subseteq A$.
This actually means that the set of associates for elements in A is
open in the set of associates. We are then led to the following definition. Note that $As(k)$ as a subset of $Tp(1)$ is a metric space.

Definition 3.8

<u>a</u> Let $\rho_k : As(k) \rightarrow Ct(k)$ be the map sending an associate for ψ

onto ψ.

<u>b</u> Let the topology T_k on $Ct(k)$ be defined by

$$A \in T_k \longleftrightarrow \rho_k^{-1}(A) \text{ is open in } As(k)$$

Theorem 3.9

Let $\varphi : Ct(k) \rightarrow \omega$. Then the following are equivalent.

<u>i</u> $\varphi \in Ct(k+1)$.

<u>ii</u> φ is continuous with respect to T_k.

Proof:
<u>i→ii</u> For each $t \in \omega$ we must show that $\varphi^{-1}\{t\}$ is open. So let
$\varphi(\psi) = t$, β an associate for ψ. β is an arbitrary element of
$\rho_k^{-1}(\varphi^{-1}\{t\})$. By Lemma 3.7 there is an n such that φ is constant t
on $B_{\beta(n)}^k$. But this means that

$$B_{\beta(n)}^1 \cap As(k) \subseteq \rho_k^{-1}(\varphi^{-1}\{t\})$$

so this set is open.

<u>ii→i</u> Assume that φ is continuous with respect to T_k. Let $\psi \in Ct(k)$
and let β be an associate for ψ. Let $t = \varphi(\psi)$. Then
$\beta \in \rho_k^{-1}(\varphi^{-1}\{t\})$, a set which by continuity of φ is open. So for some
n we have that $B_{\beta(n)}^1 \cap As(k) \subseteq \rho_k^{-1}(\varphi^{-1}\{t\})$. But then φ is constant
t on $B_{\beta(n)}^k$. By Lemma 3.7 then, $\varphi \in Ct(k+1)$. □

Remark 3.10

Using the same proof we see that $A \in T_k$ if and only if for all $\psi \in A$ and all associates β for ψ there is an n such that $B_{\bar{\beta}(n)}^k \subseteq A$.

T_k is the finest topology on $Ct(k)$ such that ρ_k is continuous, so ρ_k is what is called an <u>identification map</u> and T_k is the <u>identification topology</u>.

Since $As(k)$ is metric and T_k is an identification topology over $As(k)$ we may conclude:

Lemma 3.11

<u>a</u> T_k is generated by the convergent sequences.

<u>b</u> T_k is compactly generated.

The proofs are standard easy topological arguments and we omit them.

▫

We will put some effort in describing the convergent sequences and compact subsets of $Ct(k)$. But before we do that, we will observe a few general properties.

For instance, these spaces are what we call Lindelöf or countably compact.

Lemma 3.12

Let $A \subseteq Ct(k)$ and let $\{0_i\}_{i \in I}$ be an open covering of A. Then there is a countable subset J of I such that $\{0_i\}_{i \in J}$ covers A.

<u>Proof</u>: For each sequence-number τ, if B_τ^k is a subset of some 0_i, choose one such i to be an element of J. If $\psi \in A$, there is an i such that $\psi \in 0_i$. Let β be an associate for ψ. Then there is an n such that $B_{\bar{\beta}(n)}^k \subseteq 0_i$. Then there is a $j \in J$ such that $B_{\bar{\beta}(n)}^k \subseteq 0_j$, so $\psi \in 0_j$. This shows that $\{0_j\}_{j \in J}$ covers A and clearly J is countable.

▫

The $B_{\bar{\beta}(n)}^k$ can not be used as a basis for T_k, they are in general not open.

Lemma 3.13

Let σ be a sequence-number. Then B_σ^k is closed.

Proof: Let $\beta \in As(k)$ be an associate for ψ. We claim that

$$\psi \in B_\sigma^k \;\longleftrightarrow\; \forall n \; Con(k,\bar\beta(n),\sigma)$$

Proof of claim: Since the case $k = 1$ is trivial, we assume that $k > 1$.

$\longrightarrow$: Assume that $\psi \in B_\sigma^k$. For each n $\psi \in B_{\bar\beta(n)}^k$ so $\psi \in B_{\bar\beta(n)}^k \cap B_\sigma^k$. By definition 3.4 we see that $Con(k,\bar\beta(n),\sigma)$.

$\longleftarrow$: Now assume that $\forall n \; Con(k,\bar\beta(n),\sigma)$. We must show that σ can be extended to an associate for ψ. To do this it is sufficient to show that whenever $\sigma(\tau) = m+1$, then ψ is constant m on B_τ^{k-1}.

Assume that this is not the case, let $\varphi \in B_\tau^{k-1}$ such that $\psi(\varphi) = m' \neq m$. Let γ be an extension of τ to an associate for φ. Then there is a t such that $\beta(\bar\gamma(t)) = m'+1$. Let $n > \bar\gamma(t)$. We will show $\neg Con(k,\bar\beta(n),\sigma)$, and thereby obtain a contradiction. Assume $Con(k,\bar\beta(n),\sigma)$, and let $\psi' \in B_{\bar\beta(n)}^k \cap B_\sigma^k$. Then $\psi'(\varphi) = m'$ since $\psi' \in B_{\bar\beta(n)}^k$, $n > \bar\gamma(t)$ and $\beta(\bar\gamma(t)) = m'+1$. But $\psi'(\varphi) = m$ since ψ' has an associate extending σ, $\varphi \in B_\tau^{k-1}$ and $\sigma(\tau) = m+1$, contradicting $m \neq m'$. This ends the proof of the claim.

Now, $\{\beta; \forall n \, Con(k,\bar\beta(n),\sigma)\}$ is a closed subset of $Tp(1)$, and by the claim

$$\rho_k^{-1}(B_\sigma^k) = \{\beta; \forall n \, Con(k,\bar\beta(n),\sigma)\} \cap As(k)$$

so B_σ^k is closed.

□

Unfortunately the topology T_k is not too well behaved. For instance when $k > 1$ no points have countable pointbases, so T_k is not metrizable.

Lemma 3.14

Let $k > 1$. Let $\psi \in Ct(k)$ and let $\{0_i\}_{i \in \omega}$ be a family of open sets such that $\psi \in 0_i$ for all i. Then there is an open set 0 such that $\psi \in 0$ but no 0_i is a subset of 0.

We prove the lemma for $k = 2$. If $k > 2$ we may use the same idea, but the complexity of the associates will increase the complexity of the proof. It will be easier to formulate the proof after the more detailed analysis of the relation $Con(k,\sigma,\tau)$ in Chapter 5.

Proof when $k = 2$: Let $\psi = F \in Ct(2)$. Let β_i be an associate for F such that for all σ $\beta_i(\sigma) > 0 \rightarrow lh(\sigma) > i$. For each i choose n_i such that $B^2_{\bar\beta_i(n_i)} \subseteq 0_i$. Let γ_i be an extension of $\bar\beta_i(n_i)$ to an associate for a functional G_i such that for no σ of length $\leq i$, G_i is constant on B^1_σ , and such that $G_i \neq F$.

Clearly $G_i \in B^2_{\bar\beta_i(n)}$ so $G_i \in 0_i$.

Let $0 = Ct(2) \smallsetminus \{G_i ; i \in \omega\}$. Then $F \in 0$ and no $0_i \subseteq 0$. So it is sufficient to show that 0 is open, e.g. that $\{G_i ; i \in \omega\}$ is closed. Notice that each singleton $\{G_i\}$ is closed in T_2 since $\{G_i\} = \bigcap_{n \in \omega} B^2_{\bar\gamma_i(n)}$. It is sufficient to show that if $\{\alpha, \alpha_j\}_{j \in \omega}$ are associates, $\alpha = \lim_{j \to \infty} \alpha_j$ and each $\alpha_j \in \rho_2^{-1}\{G_i ; i \in \omega\}$ then $\alpha \in \rho_2^{-1}\{G_i ; i \in \omega\}$. For this again, it is sufficient to show that there is an m such that each $\alpha_j \in \rho_2^{-1}\{G_i ; i < m\}$.

Let σ be such that $\alpha(\sigma) > 0$. There is a j_0 such that for all $j \geq j_0$ we have $\alpha_j(\sigma) = \alpha(\sigma)$. Let $i \geq lh(\sigma)$. Then G_i is not constant on B^1_σ , so α_j cannot be an associate for G_i , i.e.

$$j \geq j_0 \rightarrow \alpha_j \in \rho_2^{-1}(\{G_i ; i \leq lh(\sigma)\})$$

It is then clear that we can find the m we are looking for and the lemma is proved for $k = 2$.

□

Remark 3.15

If $k > 2$ the push-up operator P^k_2 from Definition 1.7 is a continuous imbedding of $Ct(2)$ into $Ct(k)$. Thus, if $\psi = P^k_2(F)$ for some F , we have established the lemma for this ψ . In particular we obtain the complete corollary this way.

Corollary 3.16

If $k > 1$ then T_k is not metrizable.

□

3.2 Convergent sequences

In this section we will give two associate-free characterizations of the sequences that converge in T_k . By the following simple lemma we may use these characterizations to redefine $<Ct(k)>_{k \in \omega}$.

Lemma 3.17

Let $\phi : Ct(k) \to \omega$.

ϕ is an element of $Ct(k+1)$ if and only if for all $\{\varphi_i, \varphi\}_{i \in \omega}$ from $Ct(k)$

$$\varphi = \lim_{i \to \infty} \varphi_i \Rightarrow \phi(\varphi) = \lim_{i \to \infty} \phi(\varphi_i)$$

where $\lim_{i \to \infty} \varphi_i$ is taken in the sense of T_k .

Proof: If $\phi \in Ct(k+1)$ then ϕ is T_k-continuous, so

$$\varphi = \lim_{i \to \infty} \varphi_i \Rightarrow \phi(\varphi) = \lim_{i \to \infty} \phi(\varphi_i)$$

so assume $\forall \{\varphi_i, \varphi\}_{i \in \omega}$ from $Ct(k)$

$$\varphi = \lim_{i \to \infty} \varphi_i \Rightarrow \phi(\varphi) = \lim_{i \to \infty} \phi(\varphi_i) .$$

This means that whenever $\varphi \in \phi^{-1}\{t\}$ and $\varphi = \lim_{i \to \infty} \varphi_i$, then $\varphi_i \in \phi^{-1}\{t\}$ for almost all i . Then $\phi^{-1}\{t\}$ is open by Lemma 3.11.a. Thus ϕ is continuous.

□

Lemma 3.18

Let $k \geq 1$ and let $\{\psi_i, \psi\}_{i \in \omega}$ be elements from $Ct(k)$ such that $\psi = \lim_{i \to \infty} \psi_i$. Then, for all $\varphi \in Ct(k-1)$

$$\psi(\varphi) = \lim_{i \to \infty} \psi_i(\varphi)$$

Proof: Let $\phi(\psi') = \psi'(\varphi)$. ϕ is computable in φ so $\phi \in Ct(k+1)$. By Lemma 3.17 we have that

$$\psi(\varphi) = \phi(\psi) = \lim_{i \to \infty} \phi(\psi_i) = \lim_{i \to \infty} \psi_i(\varphi)$$

□

Remark 3.19

The converse of Lemma 3.18 is not correct, there will be sequences $\psi_i \in Ct(k)$ converging pointwise to some $\psi \in Ct(k)$ without converging to ψ in the sense of T_k . The reader may easily construct examples based on our characterizations. At this stage the argument would be too involved.

Our first characterization of the convergent sequences is not associate-free. But, imbedded in this proof we find the methods needed for our two associate-free characterizations. If $k = 1$ we are dealing with the well-known topology on $Ct(1)$ and there is no need for a characterization.

Lemma 3.20

Let $\{\psi_i, \psi\}_{i \in \omega}$ be elements of $Ct(k)$, $k > 1$. Then the following are equivalent:

<u>i</u> $\psi = \lim\limits_{i \to \infty} \psi_i$

<u>ii</u>   There are associates $\{\alpha_i, \alpha\}_{i \in \omega}$ for $\{\psi_i, \psi\}_{i \in \omega}$ resp. such that

   * $\alpha = \lim\limits_{i \to \infty} \alpha_i$

   **   For all σ and i

$$\alpha_i(\sigma) = 0 \longleftrightarrow \alpha(\sigma) = 0$$

<u>iii</u>   There are associates $\{\alpha_i, \alpha\}_{i \in \omega}$ for $\{\psi_i, \psi\}_{i \in \omega}$ resp. such that
$\alpha = \lim\limits_{i \to \infty} \alpha_i$

<u>Proof</u>: <u>ii</u> $\rightarrow$ <u>iii</u> is utterly trivial and <u>iii</u> $\rightarrow$ <u>i</u> follows from the fact
that ρ_k is continuous. So we assume <u>i</u> and want to prove <u>ii</u>.

<u>Claim</u>

Let β be an associate for an element φ in $Ct(k-1)$. Then
there is an n such that all ψ, ψ_i are constant on $B^{k-1}_{\bar\beta(n)}$.

<u>Proof</u>: Assume that the claim is false. Let n_0, t be such that ψ
is constant t on $B^{k-1}_{\bar\beta(n_0)}$. Let i_0 be such that there is an element
φ_0 with associate γ_0 such that $\bar\gamma_0(n_0) = \bar\beta(n_0)$, but $\psi_{i_0}(\varphi_0) \neq t$.
Inductively, let $n_{j+1} > n_j$ be such that for all $i \leq i_j$, ψ_i is constant
on $B^{k-1}_{\bar\beta(n_{j+1})}$. Let i_{j+1} be such that there is an element φ_{j+1} with
associate γ_{j+1} such that $\bar\gamma_{j+1}(n_{j+1}) = \bar\beta(n_{j+1})$ but $\psi_{i_{j+1}}(\varphi_{j+1}) \neq t$.
By the construction of γ_j we have that $\beta = \lim\limits_{j \to \infty} \gamma_j$, so by <u>iii</u> $\rightarrow$ <u>i</u> we
have that $\varphi = \lim\limits_{j \to \infty} \varphi_j$.
We want to conclude that $\lim\limits_{j \to \infty} \psi_{i_j}(\varphi_j) = \psi(\varphi) = t$ and thereby obtain a
contradiction. In order to do this we need:

<u>Subclaim</u>

Let $\Phi(\psi') = \mu j \, \forall j' > j \, \psi'(\varphi_{j'}) = \psi'(\varphi)$

Then Φ has an associate δ and is thus continuous.

<u>Proof</u>: Let α' be an associate for ψ'. Choose m_0 such that
$\alpha'(\bar\beta(m_0)) > 0$.
Choose m_1 such that $\forall j \geq m_1 \quad \bar\gamma_j(m_0) = \bar\beta(m_0)$.
Choose m_2 such that $\forall j < m_1 \quad (\bar\alpha'(\bar\gamma_j(m_2)) > 0)$.

Then for all j we have that $\alpha'(\bar{\gamma}_j(m_2)) > 0$. Since $\beta = \lim\limits_{j \to \infty} \gamma_j$
there are only finitely many $\bar{\gamma}_j(m_2)$ as j varies, so for some n
they are all less than n .

We let $\delta(\bar{\alpha}'(n)) = s+1$ if for all j there is an m_j and there is an
m such that $\gamma_j(m_j) < n$, $\bar{\beta}(m) < n$, $\alpha'(\gamma_j(m_j)) > 0$, $\alpha'(\bar{\beta}(m)) > 0$ and s
is minimal such that $\forall j \geq s$ $\alpha'(\bar{\gamma}_j(m_j)) = \alpha'(\bar{\beta}(m))$.

Based on the considerations above we see that δ is an associate for ϕ .

Subclaim □

Now $\psi = \lim\limits_{j \to \infty} \psi_{i_j}$, so by the subclaim there will be a j_0 such that

$$j \geq j_0 \rightarrow \phi(\psi_{i_j}) = \phi(\psi)$$

Moreover there will be a j_1 such that

$$j \geq j_1 \rightarrow \psi_{i_j}(\varphi) = \psi(\varphi) = t$$

Let $j \geq \max\{j_0, j_1, \phi(\psi)\}$. Then

$$\psi_{i_j}(\varphi_j)$$
$$= \psi_{i_j}(\varphi) \quad (\text{since } j \geq \max\{j_0, \phi(\psi)\})$$
$$= \psi(\varphi) = t \quad (\text{since } j \geq j_1)$$

This shows that $\psi(\varphi) = \lim\limits_{j \to \infty} \psi_{i_j}(\varphi_j) = t$.

But by construction $\psi_{i_j}(\varphi_j) \neq t$ so we have obtained a contradiction.

□ Claim

We are now ready to prove _i → ii_.

Define α_i by

$$\alpha_i(\bar{\beta}(n)) = \begin{cases} t+1 & \text{if all } \psi_j, \psi \text{ are constant on } B^{k-1}_{\bar{\beta}(n)} \text{ and} \\ & \psi_i \text{ is constant } t \text{ on } B^{k-1}_{\bar{\beta}(n)} \\ 0 & \text{otherwise} \end{cases}$$

α is defined similarly.

By the claim, $\{\alpha_i, \alpha\}_{i \in \omega}$ are seen to be associates for $\{\psi_i, \psi\}_{i \in \omega}$ resp.
and by construction, for all σ and for all i

$$\alpha_i(\sigma) = 0 \leftrightarrow \alpha(\sigma) = 0$$

To show that $\alpha = \lim\limits_{i \to \infty} \alpha_i$ it is sufficient to show that $\alpha(\sigma) = \lim\limits_{i \to \infty} \alpha_i(\sigma)$
for all σ such that $\alpha(\sigma) > 0$. So let $\alpha(\sigma) > 0$ and let
$\varphi \in B^{k-1}_\sigma$. Then $\alpha_i(\sigma) = 1+\psi_i(\varphi)$, $\alpha(\sigma) = 1+\psi(\varphi)$. By Lemma 3.18

$$\psi(\varphi) = \lim\limits_{i \to \infty} \psi_i(\varphi)$$

so

$$\alpha(\sigma) = \lim_{i \to \infty} \alpha_i(\sigma)$$

This ends the proof of the lemma.

◻

Remark 3.21

We did not describe $\alpha_i(\sigma)$ for those σ such that $\bar{B}_\sigma^{k-1} = \emptyset$. As a convention we may let $\alpha_i(\sigma) = \alpha_i(\bar{\sigma}(t))$, where $0 \leq t < \mathrm{lh}(\sigma)$ is maximal such that $\bar{\sigma}(t)$ can be extended to an associate. Such σ's are, however, of no importance and later in similar situations we will forget them completely.

Lemma 3.20 and its proof enables us to give an associate-free description of the convergent sequences. Our first theorem is due to Martin Hyland [20].

Theorem 3.22

Let $k > 0$, $\{\psi_i, \psi\}_{i \in \omega}$ be elements of $Ct(k)$. Then the following are equivalent:

<u>i</u> $\psi = \lim_{i \to \infty} \psi_i$

<u>ii</u>   For all $\{\varphi_i, \varphi\}_{i \in \omega}$ from $Ct(k-1)$

$$\varphi = \lim_{i \to \infty} \varphi_i \;\Rightarrow\; \psi(\varphi) = \lim_{i \to \infty} \psi_i(\varphi_i)$$

Proof:

<u>i</u> → <u>ii</u> Let $\psi = \lim_{i \to \infty} \psi_i$, $\varphi = \lim_{i \to \infty} \varphi_i$. By <u>iii</u> of Lemma 3.20, let $\{\alpha_i, \alpha\}_{i \in \omega}$ and $\{\beta_i, \beta\}_{i \in \omega}$ be associates for $\{\psi_i, \psi\}_{i \in \omega}$ and $\{\varphi_i, \varphi\}_{i \in \omega}$ resp. such that $\alpha = \lim_{i \to \infty} \alpha_i$ and $\beta = \lim_{i \to \infty} \beta_i$ Choose n such that $\alpha(\bar{\beta}(n)) > 0$. Let i_0 be such that $i \geq i_0 \Rightarrow \bar{\beta}_i(n) = \bar{\beta}(n)$. Let i_1 be such that $i \geq i_1 \Rightarrow \alpha_i(\bar{\beta}(n)) = \alpha(\bar{\beta}(n))$. Then for $i \geq \max\{i_0, i_1\}$ we have that

$$\alpha_i(\bar{\beta}_i(n)) = \alpha_i(\bar{\beta}(n)) = \alpha(\bar{\beta}(n))$$

so

$$\psi_i(\varphi_i) = \psi(\varphi) \quad \text{for such } i .$$

<u>ii</u> → <u>i</u> Here we repeat the argument of <u>i</u> → <u>ii</u> from the proof of Lemma 3.20, using assumption 3.22 <u>ii</u> instead of the subclaim.

◻

This theorem is the basis for Martin Hyland's limit-space interpretation of the countable functionals. We will define the basic notion
of a limit-space in order to formulate his characterization. For further
information about limit-spaces see Hyland [20] or Kuratowski [25].

Definition 3.23

a   A <u>limit-space</u> $<L,\to>$ is a set L with a partial map $\to$ from the
set of sequences from L to L satisfying:

  <u>i</u> If $\{a_i\}_{i\in\omega}$ is a sequence such that $a_i = a$ for almost all
    $i \in \omega$, then $\{a_i\}_{i\in\omega} \to a$.

  <u>ii</u> If $\{a_i\}_{i\in\omega} \to a$ and $\{a_{i_j}\}_{j\in\omega}$ is a subsequence of $\{a_i\}_{i\in\omega}$
    then $\{a_{i_j}\}_{j\in\omega} \to a$.

  <u>iii</u> If $\{a_i\}_{i\in\omega} \not\to a$ there is a subsequence $\{a_{i_j}\}_{j\in\omega}$ such that for
    no subsequence $\{a_{i_{j_k}}\}_{k\in\omega}$ of $\{a_{i_j}\}_{j\in\omega}$
    $\{a_{i_{j_k}}\}_{k\in\omega} \to a$.

<u>b</u>   If $<L_1,\to>$ and $<L_2,\to>$ are two limit-spaces and $\phi : L_1 \to L_2$ then
ϕ is said to be <u>continuous</u> if

$$\{a_i\}_{i\in\omega} \to a \Rightarrow \{\phi(a_i)\}_{i\in\omega} \to \phi(a)$$

for all $\{a_i, a\}_{i\in\omega}$ from L_1.

<u>c</u>   The cartesian product of two limit-spaces L_1 and L_2 is organ-
ized to a limit-space by

$$\{(a_i, b_i)\}_{i\in\omega} \to (a,b) \text{ if } \{a_i\}_{i\in\omega} \to a \;\&\; \{b_i\}_{i\in\omega} \to b$$

<u>d</u>   The set of continuous functions from one limit-space $<L_1,\to>$ to
another limit-space $<L_2,\to>$ is organized to a limit-space by

$$\{\phi_i\}_{i\in\omega} \to \phi$$

if

$$\{a_i\}_{i\in\omega} \to a \text{ in } L_1 \Rightarrow \{\phi_i(a_i)\}_{i\in\omega} \to \phi(a) \text{ in } L_2.$$

<u>e</u>   The natural numbers is organized to a limit-space by

$$\{n_i\}_{i\in\omega} \to n \text{ if } n_i = n \text{ for almost all } i.$$

Remark 3.24

It is an easy exercise to check that these constructions will give
us limit-spaces.

As a consequence of Theorem 3.22 we then obtain

Corollary 3.25

Let $L_0 = \langle\omega,\to\rangle$ as defined in 3.23 $\underline{e}$.
Inductively let $\langle L_{k+1},\to\rangle$ be the limit-space of continuous functions
from $\langle L_k,\to\rangle$ to $\langle\omega,\to\rangle$. Then for all k we have that $L_k = Ct(k)$
and T_k is the topology generated from $\to$ on L_k.

We have defined the topologies T_k only on the spaces $Ct(k)$ of
pure type, while we defined the limit structure also on cartesian products

$$Ct(k_1) \times \ldots \times Ct(k_n)$$

Clearly our computable functions are maps from such cartesian products
into ω, and if we are going to define a topology on $Ct(k_1)\times\ldots\times Ct(k_n)$
we would like to do it in such a way that all computable functionals of
mixed type are continuous. As we will see in Lemma 3.27 we cannot use
the product topology on $Ct(k_1)\times\ldots\times Ct(k_n)$. At this stage we are then
left with two alternatives:

$\underline{i}$ $0 \subseteq Ct(k_1)\times\ldots\times Ct(k_n)$ is open iff $(\rho_{k_1}\times\ldots\times\rho_{k_n})^{-1}(0)$ is open in
     $As(k_1)\times\ldots\times As(k_n)$.

$\underline{ii}$ $0 \subseteq Ct(k_1)\times\ldots\times Ct(k_n)$ is open if whenever $\{\vec{\phi}_i\}_{i\in\omega} \to \vec{\phi}$ in the
     limit-space sense, and $\vec{\phi} \in 0$, then $\vec{\phi}_i \in 0$ for almost all $i \in \omega$.

As a consequence of Theorem 2.28 we see that all computable functionals will be continuous if we use alternative $\underline{i}$. The two alternatives are in fact equivalent.

Lemma 3.26

$\underline{i}$ and $\underline{ii}$ above are equivalent.

Proof:

$\underline{i} \to \underline{ii}$ Let $0 \subseteq Ct(k_1)\times\ldots\times Ct(k_n)$ be open by $\underline{i}$.
     Let $\vec{\phi} = (\phi^1,\ldots,\phi^n) \in 0$ and let $\vec{\phi}_i = (\phi_i^1,\ldots,\phi_i^n)$ be such
that $\{\vec{\phi}_i\}_{i\in\omega} \to \vec{\phi}$. Then for each $j \leq n$, $\phi^j = \lim_{i\to\infty} \phi_i^j$.
By Lemma 3.20 part $\underline{iii}$ let $\{a_i^j,a^j\}_{i\in\omega,j\leq n}$ be associates for
$\{\phi_i^j,\phi^j\}_{i\in\omega,j\leq n}$ resp. such that $a^j = \lim_{i\to\infty} a_i^j$. Then $\vec{a} = \lim_{i\to\infty} \vec{a}_i$, and
since $\vec{a} \in (\rho_{k_1}\times\ldots\times\rho_{k_n})^{-1}(0)$, which is open, we must have that
$\vec{a}_i \in (\rho_{k_1}\times\ldots\times\rho_{k_n})^{-1}(0)$ for almost all $i \in \omega$.
It follows that $\vec{\phi}_i \in 0$ for almost all $i \in \omega$.

$\underline{ii} \to \underline{i}$ Assume that 0 satisfies $\underline{ii}$ and let $B = (\rho_{k_1} \times \ldots \times \rho_{k_n})^{-1}(0)$,

Let $\vec{\alpha} \in B$ and let $\{\vec{\alpha}_i\}_{i \in \omega}$ be sequences of associates such that $\vec{\alpha} = \lim\limits_{i \to \infty} \vec{\alpha}_i$. Let $\psi_i^j = \rho_{k_j}(\alpha_i^j)$ and $\psi^j = \rho_{k_j}(\alpha^j)$ $(i \in \omega, \ j \leq n)$. Then $\psi^j = \lim\limits_{i \to \infty} \psi_i^j$, so $\{\vec{\psi}_i\}_{i \in \omega} \to \vec{\psi}$. By $\underline{ii}$, $\vec{\psi}_i \in 0$ for almost all $i \in \omega$. But then $\vec{\alpha}_i \in B$ for almost all $i \in \omega$.

This shows that B is open in $As(k_1) \times \ldots \times As(k_n)$, so 0 is open by $\underline{i}$.

□

Lemma 3.27

Let $Ev(F,f) = F(f)$ where Ev is defined on $Ct(2) \times Ct(1)$. Then Ev is not continuous in the product topology.

Proof: Let F, f be given, $F(f) = t$.

Claim

For each n there is a sequence $F_i \to F$ such that for each i there is a $g \in B_{\bar{f}(n)}^1$ such that $F_i(g) = t+1$.

From the claim we show that $Ev^{-1}\{t\}$ is not open in the product topology by the following argument:

If $Ev^{-1}\{t\}$ is open in the product topology there is an open set $0 \subseteq Ct(2)$ and an n such that $F \in 0$ and $0 \times B_{\bar{f}(n)}^1 \subseteq Ev^{-1}\{t\}$. Let $\{F_i\}_{i \in \omega} \to F$ be as in the claim. For some i , $F_i \in 0$ and then by the claim there is a $g \in B_{\bar{f}(n)}^1$ such that $Ev(F_i,g) = t+1$. But $(F_i,g) \in 0 \times B_{\bar{f}(n)}^1$, so $Ev(F_i,g) = t$, contradiction.

Proof of claim: Let β be an associate for F such that for all σ , $\beta(\sigma) > 0 \to lh(\sigma) > n$.
Let

$$\beta_i(\sigma) = \begin{cases} \beta(\sigma) & \text{if } \beta(\sigma) = 0 \text{ or } \sigma(n) \neq i \\ t+2 & \text{if } \beta(\sigma) > 0 \text{ and } \sigma(n) = i \end{cases}$$

Now β_i is an associate for the functional

$$F_i(g) = \begin{cases} F(g) & \text{if } g(n) \neq i \\ t+1 & \text{if } g(n) = i \end{cases}$$

$\beta = \lim\limits_{i \to \infty} \beta_i$ so $F = \lim\limits_{i \to \infty} F_i$ by lemma 3.20 iii.
Moreover, if

$$g_i(m) = \begin{cases} f(m) & \text{if } m \neq n \\ i & \text{if } m = n \end{cases}$$

we see that $g_i \in B^1_{\frac{1}{f}(n)}$ and $F_i(g_i) = t+1$.
This ends the proof of the claim and the lemma.

□

We will now turn to another characterization of the convergent sequences.

Definition 3.28
    Let $\{\psi_i\}_{i\in\omega}$ be a sequence from $Ct(k)$, $k > 0$. $\phi \in Ct(k)$ is called a modulus functional for $\{\psi_i\}_{i\in\omega}$ if

$$(\forall\varphi \in Ct(k-1))(\forall i,j \geq \phi(\varphi))(\psi_i(\varphi) = \psi_j(\varphi))$$

Remark 3.29
    That ϕ is a modulus functional for $\{\psi_i\}_{i\in\omega}$ means that for each φ $\psi_i(\varphi)$ will converge, and $\phi(\varphi)$ computes an index i after which $\psi_j(\varphi)$ is constant, i.e. the limit value is obtained.

Theorem 3.30
    Let $k \geq 1$ and let $\{\psi_i\}_{i\in\omega}$ be a sequence from $Ct(k)$. Then the following are equivalent

i $\lim_{i\to\infty} \psi_i$ exists in $Ct(k)$

ii $\{\psi_i\}_{i\in\omega}$ has a modulus functional ϕ in $Ct(k)$.

Proof: For $k = 1$ this is trivial, so assume $k > 1$.

ii → i Let ϕ be a modulus functional for $\{\psi_i\}_{i\in\omega}$. Then the point-wise limit ψ is computable in ϕ and $<\psi_i>_{i\in\omega}$ by

$$\psi(\varphi) = \psi_{\phi(\varphi)}(\varphi)$$

so ψ is countable.
We will use Theorem 3.22 to show that $\psi = \lim_{i\to\infty} \psi_i$. So let $\{\varphi_i,\varphi\}_{i\in\omega}$ be elements of $Ct(k-1)$ such that $\varphi = \lim_{i\to\infty} \varphi_i$.
It is sufficient by 3.22 to show that $\psi(\varphi) = \lim_{i\to\infty} \psi_i(\varphi_i)$. Let i_0 be such that $i \geq i_0 \to \phi(\varphi_i) = \phi(\varphi)$. Let $j = \phi(\varphi)$ and let i_1 be such that $i \geq i_1 \to \psi(\varphi_i) = \psi(\varphi)$. Then

$i \geq \max\{i_0,i_1,j\} \to \phi(\varphi_i) = j$ (since $i \geq i_0$) so

$i \geq \max\{i_0,i_1,j\} \to \psi_i(\varphi_i) = \psi(\varphi_i)$ (since $i \geq \phi(\varphi_i)$)

$\qquad\qquad\qquad\qquad = \psi(\varphi)$ (since $i \geq i_1$)

This shows that $\psi(\varphi) = \lim_{i\to\infty} \psi_i(\varphi_i)$ and ii → i is proved.

<u>i →ii</u> Assume that $\psi = \lim\limits_{i\to\infty} \psi_i$. By Lemma 3.20 <u>ii</u> there are associates

$\{\alpha_i,\alpha\}_{i\epsilon\omega}$ for $\{\psi_i,\psi\}_{i\epsilon\omega}$ resp. such that

* $\forall\sigma \; \forall i \; \alpha_i(\sigma) = 0 \longleftrightarrow \alpha(\sigma) = 0$

** $\alpha = \lim\limits_{i\to\infty} \alpha_i$.

Let

$$\beta(\sigma) = \begin{cases} 0 & \text{if } \alpha(\sigma) = 0 \\ t+1 & \text{if } \alpha(\sigma) > 0 \text{ and } t \text{ is minimal such that} \\ & \forall i \geq t \; \alpha_i(\sigma) = \alpha(\sigma) \end{cases}$$

β will then be an associate for a modulus functional ϕ for $\{\psi_i\}_{i\epsilon\omega}$.

This theorem permits us to generalize Theorem 2.24. The generalization is due to J. Bergstra [3]. Our next result also gives a recursion-theoretic characterization of $Ct(k)$.

<u>Theorem 3.31</u>

Let $\Psi:Ct(k) \to \omega$. Then the following are equivalent:

<u>i</u> $\Psi \in Ct(k+1)$

<u>ii</u> ^2E is not computable in Ψ and any element of $Ct(k)$.

<u>Proof</u>:

<u>i →ii</u> is trivial since ^2E is not countable and the countable
      functionals are closed under computations.

<u>ii →i</u> Assume that $\Psi \in Ct(k+1)$. By Lemma 3.17 there is a sequence
      $\{\psi_i,\psi\}_{i\epsilon\omega}$ from $Ct(k)$ such that $\psi = \lim\limits_{i\to\infty} \psi_i$ but
$\neg(\Psi(\psi) = \lim\limits_{i\to\infty} \Psi(\psi_i))$, w.l.o.g. we may assume for all i
$\Psi(\psi) \neq \Psi(\psi_i)$, and even that $\Psi(\psi) = 1$ while $\Psi(\psi_i) = 0$ for all i . By
Theorem 3.30, let ϕ be a modulus functional for $\{\psi_i\}_{i\epsilon\omega}$. We will
give an algorithm for computing $^2E(f)$ from f, Ψ, ϕ, ψ and $<\psi_i>_{i\epsilon\omega}$.

Let $\gamma_f(\varphi)$ be defined by:

If $\exists n \leq \phi(\varphi) \; (f(n) = 0)$ let $\gamma_f(\varphi) = \psi_n(\varphi)$ for the least such n .

If $\forall n \leq \phi(\varphi) \; (f(n) \neq 0)$ let $\gamma_f(\varphi) = \psi_{\phi(\varphi)}(\varphi) = \psi(\varphi) = \psi_m(\varphi)$ for all
$$m \geq \phi(\varphi) .$$

It is then easy to see that

$\exists n(f(n) = 0) \to \gamma_f = \psi_n$ for the least such n

$\forall n(f(n) \neq 0) \to \gamma_f = \psi$.

Moreover, γ_f is computable in the given parameters. We then have $\exists n\ f(n) = 0 \leftrightarrow \exists n\ \gamma_f = \psi_n \leftrightarrow \Psi(\gamma_f) = 1$. So $^2E(f) = \Psi(\gamma_f)$.
This shows that ^2E is computable in the given parameters.

□

Remark 3.32

This theorem can also be proved directly on the basis of Theorem 3.22, as shown in Hyland [20]. We have given this version since the modulus functionals will be used quite a lot later, and we will meet this trick again. The particular trick of this argument is essentially due to Grilliot [14].

3.3 Compact sets in Ct(k)

In this section we will give a series of characterizations of the compact subsets of Ct(k) . These are of interest since T_k , being an identification-topology over a metric space, is compactly generated. We will use these characterizations to show that the topological spaces $<Ct(k),T_k>$ may be obtained through standard topological constructions of function spaces.

For $<Ct(1),T_1>$ the situation is quite familiar:

A set $A \subseteq Ct(1)$ is <u>bounded</u> if there is a function f such that $\forall g \in A\ \forall n\ g(n) \leq f(n)$.

$C \subseteq Ct(1)$ is compact if and only if C is closed and bounded.

We may try to lift this result to higher types, and we will need the following definition.

Definition 3.33

<u>a</u>   Let $A \subseteq Ct(k)$. A is said to be <u>bounded</u> if for each $\varphi \in Ct(k-1)$ there is an n such that $\psi \in A \rightarrow \psi(\varphi) \leq n$.

<u>b</u> $A \subseteq Ct(k)$ is said to be <u>continuously bounded</u> if there is a $\Phi \in Ct(k)$ such that

$$\forall \varphi \in Ct(k-1)\ \forall \psi \in A\ (\psi(\varphi) \leq \Phi(\varphi))$$

Lemma 3.34

If $C \subseteq Ct(k)$ is compact, then C is closed and bounded.

<u>Proof</u>: Let $\varphi \in Ct(k-1)$ be given. Let $0_t = \{\psi; \psi(\varphi) = t\}$. $\{0_t\}_{t \in \omega}$ is

an open covering of $Ct(k)$, so by compactness of C there will be an n such that $C \subseteq O_0 U...U O_n$. But then $\forall \psi \in C (\psi(\varphi) \leq n)$, so C is bounded.

$Ct(k)$ is a Hausdorff-space so all compact sets are closed. □

Remark 3.35

We will later show that a compact set C will be continuously bounded, but we will first consider an example showing that there are closed continuously bounded sets that are not compact.

Lemma 3.36

Let $D = \{G \in Ct(2); \forall f \in Ct(1) G(f) \leq 1\}$. Then D is continuously bounded and closed but not compact.

Proof: D is bounded by the constant functional 1. To show that D is closed, we see that if $\alpha \in As(2)$ then

$$\rho_2(\alpha) \in D \longleftrightarrow \forall \sigma (\alpha(\sigma) \leq 2)$$

Now $\{\alpha: \forall \sigma \alpha(\sigma) \leq 2\}$ is closed, even compact, so $\rho_2^{-1}(D)$ is closed in $As(2)$.

In order to show that D is not compact we show that $\mathbb{N} = \{\Phi(G,g); G \in D\}$ where Φ is the fan functional and g is constant 1. (For the definition of Φ see chapter 4.)

So let $n \in \mathbb{N}$. If $n = 0$ we let $G(f) = 0$ for all f. Then $\Phi(G,g) = 0$ and $G \in D$.
If $n > 0$ we let

$$G(f) = \begin{cases} 0 & \text{if } f(n-1) = 0 \\ 1 & \text{if } f(n-1) \neq 0 \end{cases}$$

Then $G \in D$ and $\Phi(G,g) = n$.

But $\lambda G \Phi(G,g)$ is continuous and the continuous image of a compact set is compact. $\mathbb{N}$ is not compact, so D cannot be compact. □

Remark 3.37

It is not the case that $\rho_2^{-1}(D)$ is compact in $As(2)$ although it is the intersection of $As(2)$ with a compact set.

An important property of convergent sequences is that we can find associates for the elements of the sequence having the same zero's. When we investigate the compact subsets of $Ct(k)$ this property will be so important that we give it a name.

Definition 3.38

Let $k \geq 2$, $A \subseteq Ct(k)$. A is said to be _equicontinuous_ if there are associates α_ψ for each element ψ of A such that

$$\forall \sigma \; \forall \psi, \psi' \in A \; (\alpha_\psi(\sigma) > 0 \Rightarrow \alpha_{\psi'}(\sigma) > 0)$$

Remark 3.39

a $Ct(1)$ is a metric space, so the notion of equicontinuity of subsets of $Ct(2)$ is defined in the literature. It is an easy exercise to show that our notion and the standard one coincide for subsets of $Ct(2)$. (We may also use Lemmas 3.40 and 3.41 for this.)

b Let $\psi \in Ct(k)$, $\varphi \in Ct(k-1)$. A measure of continuity of ψ at φ is a function that to any associate β for φ gives an n such that ψ is constant on $B_{\beta(n)}^{k-1}$. So our notion coincides with this measure of continuity.

In the next two lemmas we will show that our notion of equicontinuity coincides with a definition independent of the associates. These lemmas are a digression and will not be used later in the text.

Lemma 3.40

Let $k \geq 2$. Let $A \subseteq Ct(k)$ be equicontinuous, $\varphi \in Ct(k-1)$. Then there is an open set $0 \subseteq Ct(k-1)$ such that $\varphi \in 0$ and

$$\forall \psi \in A \; (\psi \text{ is constant on } 0)$$

Proof: Let $\Sigma = \{\sigma; \text{ all } \psi \in A \text{ are constant on } B_\sigma^{k-1}\}$. Let β be an associate for φ. Since A is equicontinuous there will be an n such that $\bar{\beta}(n) \in \Sigma$.
Let $\sigma \in E \iff \exists m \; \exists(\sigma_0, \ldots, \sigma_m)$ (Each $\sigma_i \in \Sigma$ $(i \leq m)$,

$$\text{Con}(k-1, \sigma_i, \sigma_{i+1}) \; (i \leq m-1),$$

$$\sigma_0 = \bar{\beta}(n) \text{ for some } n \text{ and } \sigma_m = \sigma.)$$

(The relation $\text{Con}(k, \sigma, \tau)$ is defined in 3.4.)
Let $B = \bigcup_{\sigma \in E} B_\sigma^1$.

B is open in $Ct(1)$ since it is a union of basis elements.
The lemma will then follow from:

1. $B \cap As(k-1)$ is the ρ_{k-1}-inverse of some set 0 in $Ct(k-1)$ (which then is open by the definition of T_{k-1}).

2. Each $\psi \in A$ is constant $\psi(\varphi)$ on 0, and $\varphi \in 0$.

Proof of 1: Let α_1, α_2 be associates for the same functional φ' and assume that $\alpha_1 \in B$. We will show that $\alpha_2 \in B$. By definition of B there is an n such that $\bar{\alpha}_1(n) \in E$. Choose $(\sigma_0, \ldots, \sigma_m)$ as in the definition of $\bar{\alpha}_1(n) \in E$. By equicontinuity of A there will be some n' such that $\bar{\alpha}_2(n') \in \Sigma$. Since $\varphi' \in B_{\bar{\alpha}_1(n)}^{k-1} \cap B_{\bar{\alpha}_2(n')}^{k-1}$ we have $\mathrm{Con}(k-1, \bar{\alpha}_1(n), \bar{\alpha}_2(n'))$. By regarding the sequence $(\sigma_0, \ldots, \sigma_m, \bar{\alpha}_2(n'))$ we see that $\bar{\alpha}_2(n') \in E$. But then $\alpha_2 \in B$.

Let $\varphi' \in 0 \longleftrightarrow \varphi'$ has an associate in B. Then $B \cap As(k-1) = \rho_{k-1}^{-1}(0)$ by the argument above.

Proof of 2: Let $\sigma \in E$ and choose $(\sigma_0, \ldots, \sigma_m)$ according to the definition of E. By induction on m we prove that for any $\psi \in A$, ψ is constant $\psi(\varphi)$ on $B_{\sigma_m}^{k-1}$. The induction is trivial, and it follows that for any $\psi \in A$, ψ is constant on 0 defined in 1. Clearly $\varphi \in 0$.

□

The converse is also true:

Lemma 3.41

Let $k \geq 2$, $A \subseteq Ct(k)$. Assume that for each $\varphi \in Ct(k-1)$ there is an open set $0 \subseteq Ct(k-1)$ such that $\varphi \in 0$ and each $\psi \in A$ is constant on 0. Then A is equicontinuous.

Proof: Let φ be given, β Let 0 be given according to the assumption of the lemma. Since $\varphi \in 0$ there is an n such that $B_{\bar{\beta}(n)}^{k-1} \subseteq 0$. So all $\psi \in A$ are constant on $B_{\bar{\beta}(n)}^{k-1}$.

From the fact that for each $\beta \in As(k-1)$ there is an n such that all $\psi \in A$ are constant on $B_{\bar{\beta}(n)}^{k-1}$ it is easy to construct associates α_ψ for the elements ψ of A such that they all have the same zeros.

□

After this digression, which serves as a justification for our choice of the term 'equicontinuous' we proceed along the main track:

Lemma 3.42

Let $k \geq 2$ and let $A \subseteq Ct(k)$. If A is equicontinuous and bounded, then A is continuously bounded.

Proof: Let $\{\alpha_\psi ; \psi \in A\}$ be associates for the elements of A with the same zeros. Let

$$\alpha(\sigma) = \begin{cases} 0 & \text{if some } \alpha_\psi(\sigma) = 0 \ (\longleftrightarrow \text{ all } \alpha_\psi(\sigma) = 0) \\ \max\{\alpha_\psi(\sigma) ; \psi \in A\} & \text{if each } \alpha_\psi(\sigma) > 0 \end{cases}$$

Then α will be an associate for the functional

$$\Phi(\varphi) = \max\{\psi(\varphi); \psi \in A\}$$

□

Lemma 3.43

Let $k \geq 2$. Let $C \subseteq Ct(k)$ be compact.
Then C is equicontinuous.

Proof: Assume that C is not equicontinuous. Then there must be a
$\varphi \in Ct(k-1)$ and an associate β for φ such that for each n there
is a $\psi_n \in C$ such that ψ_n is not constant on $B_{\bar{\beta}(n)}^{k-1}$.
Since C is compact, $\{\psi_n\}_{n\in\omega}$ has a convergent subsequence
$\{\psi_{n_j}\}_{j\in\omega}$. This sequence cannot be equicontinuous, by choice of ψ_n.
But by Lemma 3.20 ii all convergent sequences are equicontinuous, a
contradiction.
□

Lemma 3.44

Let $k \geq 2$. Let $C \subseteq Ct(k)$ be closed, bounded and equicontinuous
Then C is compact.

Proof: By Lemma 3.42 C is continuously bounded and by the proof of
that lemma we may find a bound Φ with an associate α_0 such that

$$\forall\psi \in C \ (\alpha_0(\sigma)>0 \to \psi \text{ is constant on } B_\sigma^{k-1})$$

Let $\alpha \in C_0$ if $\forall\sigma \ (\alpha(\sigma) = 0 \leftrightarrow \alpha_0(\sigma) = 0)$

$$\wedge \ \forall\sigma \ (\alpha(\sigma) \leq \alpha_0(\sigma))$$

$$\wedge \ \forall\sigma,\tau \ (Con(k-1,\sigma,\tau) \wedge \alpha(\sigma)>0 \wedge \alpha(\tau)>0$$

$$\to \alpha(\sigma) = \alpha(\tau))$$

C_0 is a closed subset of the compact set $C_{\alpha_0} = \{\alpha : \forall n \ \alpha(n) \leq \alpha_0(n)\}$
so C_0 is compact.

Claim 1

If $\alpha \in C_0$ then $\alpha \in As(k)$.

Proof: Let $\beta \in As(k-1)$, $\alpha \in C_0$. Since $\alpha_0 \in As(k)$ there is an n
such that $\alpha_0(\bar{\beta}(n)) > 0$, and by definition of C_0, $\alpha(\bar{\beta}(n)) > 0$ for all
such n.
If β_1 and β_2 are associates for the same functional, $\alpha_0(\bar{\beta}_1(n)) > 0$
and $\alpha(\bar{\beta}_2(m)) > 0$, then we know that $Con(k-1,\bar{\beta}_1(n),\bar{\beta}_2(m))$, so by the

last part of the definition of C_0 we see that $\alpha(\bar{\beta}_1(n)) = \alpha(\bar{\beta}_2(m))$. But then $\alpha \in As(k)$.

$\square$ Claim 1

Claim 2

If $\psi \in C$ then ψ has an associate in C_0.

Proof:

Let $\alpha(\sigma) = \begin{cases} 0 & \text{if } \alpha_0(\sigma) = 0 \\ t+1 & \text{if } \alpha_0(\sigma) > 0 \text{ and } \psi \text{ is constant } t \text{ on } B_\sigma^{k-1} \end{cases}$

By choice of α_0 we see that α is an associate for ψ and clearly $\alpha \in C_0$.

$\square$ Claim 2

Since C is closed there is a closed set $K \subseteq Ct(1)$ such that $\rho_k^{-1}(C) = K \cap As(k)$. Let $C_1 = K \cap C_0$. Then C_1 is compact, each element of C_1 is an associate for an element of C and each $\psi \in C$ has an associate in C_1. (This is seen from the claims and the definitions of K and C_1.) Thus $C = \rho_k'' C_1$, so C is the continuous image of a compact set, and thus C is itself compact.

$\square$

With these lemmas and their proofs we have sufficient information to state

Theorem 3.45

Let $k \geq 2$ and let $C \subseteq Ct(k)$. Then the following are equivalent

i C is compact

ii C is closed, bounded and equicontinuous

iii C has a compact set of associates, i.e. $C = \rho_k'' C_1$ for some compact $C_1 \subseteq As(k)$

iv C is homeomorphic to a compact subset of $Ct(1)$

Proofs:

i → ii Lemma 3.34 and Lemma 3.43

ii → iii This was established in the proof of Lemma 3.44

iii → i Trivial, as in the proof of Lemma 3.44

iv → i Trivial

ii → iv We notice that C_1 constructed in the proof of Lemma 3.44

contains exactly one associate for each $\psi \in C$, so $\rho_k \upharpoonright C_1$ is one to one. By standard elementary topology $\rho_k \upharpoonright C_1$ is then a homeomorphism.

□

Topologists investigate various interesting topologies on a function-space. One of these standard constructions is of interest to us, namely the compact-open topology.

Definition 3.46

Let $<X,T_1>$ and $<Y,T_2>$ be two topological spaces. Let

$$Z = [X,Y] = \{F : X \to Y ; F \text{ is continuous}\}$$

The compact-open topology on Z is defined by using the following basis:

Finite intersections of sets of the form

$$Z_{[C,0]} = \{F ; F"C \subseteq 0\}$$

where $C \subseteq X$ is compact and $0 \subseteq Y$ is open.

In what remains of this section we let $k \geq 1$. We will investigate the compact open topology on $Ct(k+1) = [Ct(k),\omega]$.

A typical element of the basis for the compact-open topology on $Ct(k+1)$ will be a finite intersection of sets of the form

$$0_{C,A} = \{\psi ; \psi"C \subseteq A\}$$

where C is compact and $A \subseteq \omega$.

We will first find a more simple basis.

Lemma 3.47

The class

$$\{0_{C_1,\{n_1\}} \cap \ldots \cap 0_{C_t,\{n_t\}} ; C_1,\ldots,C_t \text{ are compact subsets of}$$
$$Ct(k) \text{ and } n_1,\ldots,n_t \in \omega\}$$

is a basis for the compact-open topology on $Ct(k+1)$.

Proof: These sets are clearly open in the compact-open topology.

Let $\varphi \in 0_{C_1,A_1} \cap \ldots \cap 0_{C_t,A_t}$ where $C_1,\ldots,C_t$ are compact and $A_1,\ldots,A_t$ are subsets of ω . For each $i \leq t$ there will be a finite partition of C_i into compact subsets $C_{i,1},\ldots,C_{i,n_i}$ and values $m_{i,1},\ldots,m_{i,n_i}$ in A_i such that φ is constant $m_{i,j}$ on $C_{i,j}$. (Here we use that C_i is compact and that $\varphi^{-1}\{m_{i,j}\}$ is both closed and open.) Then

$$\varphi \in {}^0C_{1,1},\{m_1,1\}\cap \ldots \cap {}^0C_{1,n_1},\{m_1,n_1\}\cap \ldots \cap {}^0C_t,n_t,\{m_t,n_t\}$$

$$\subseteq {}^0C_1,A_1 \cap \ldots \cap {}^0C_t,A_t \qquad \square$$

<u>Lemma 3.48</u>

If $0 \subseteq Ct(k+1)$ is open in the compact-open topology, then 0 is open in T_{k+1} .

<u>Proof</u>: By Lemma 3.47 it is sufficient to show that each ${}^0C,\{n\}$ is open in T_{k+1} .

So, let $\psi \in {}^0C,\{n\}$, i.e. ψ is constant n on C . Let α be an associate for ψ . By the definition of T_{k+1} it is sufficient to find an m such that for all $\psi' \in B^{k+1}_{\bar{\alpha}(m)}$, ψ' is constant n on C .

Let C_1 be a compact set of associates such that $C = \rho''_{k+1}C_1$ (Theorem 3.45 <u>iii</u>). For each $\beta \in C_1$ there is a t such that $\alpha(\bar{\beta}(t)) = n+1$.

The map $\beta \rightarrow \mu t\, \alpha(\bar{\beta}(t)) = n+1$ is continuous on C_1 and thus bounded by some t_0 since C_1 is compact. Again, since C_1 is compact $\{\bar{\beta}(t_0); \beta \in C_1\}$ is finite. So choose m so large that

$$\forall \beta \in C_1 \exists t\, (\bar{\beta}(t) < m \wedge \alpha(\bar{\beta}(t)) = n+1)$$

If ψ' has an associate α' extending $\bar{\alpha}(m)$, then $\alpha'(\bar{\beta}(t)) = n+1$ whenever $\beta \in C_1$ and $\alpha'(\bar{\beta}(t)) > 0$. So ψ' will be constant n on C and $B^{k+1}_{\bar{\alpha}(m)} \subseteq {}^0C,\{n\}$. $\qquad \square$

By Lemma 3.48 the compact-open topology on $Ct(k+1)$ is a coarser topology than T_{k+1} , so all compact subsets of $Ct(k+1)$ are compact in the compact-open topology.

We will show that the compact sets are the same.

<u>Lemma 3.49</u>

Let $C \subseteq Ct(k+1)$ be compact in the compact-open topology. Then C is compact in T_{k+1} .

<u>Proof</u>: We will show that C is closed, bounded and equicontinuous and then use Theorem 3.45 <u>ii</u> $\rightarrow$ <u>i</u>.

<u>C is closed</u> in T_{k+1} since C is closed in the compact-open topology

<u>C is bounded</u>: Let $\varphi \in Ct(k)$. Let $\Psi(\psi) = \psi(\varphi)$. Ψ is constant n on

the sets $0_{\{\varphi\},\{n\}}$ which are open in the compact-open topology, so Ψ is continuous in the compact-open topology.
Then

$$\{\Psi(\psi); \psi \in C\} = \{\psi(\varphi); \psi \in C\}$$

is bounded since C is compact.

C is equicontinuous: We want to repeat the argument of 3.43. In order to do so it is sufficient to show

Claim

Let $\{\psi_i, \psi\}_{i \in \omega}$ be elements of $Ct(k+1)$. If $\psi = \lim\limits_{i \to \infty} \psi_i$ in the compact-open topology, then $\psi = \lim\limits_{i \to \infty} \psi_i$ in T_{k+1}.

Proof: Assume $\psi = \lim\limits_{i \to \infty} \psi_i$ in the compact-open topology. We are going to use Theorem 3.22, so let $\{\varphi_i, \varphi\}_{i \in \omega}$ be elements of $Ct(k)$ such that $\varphi = \lim\limits_{i \to \infty} \varphi_i$. We may assume w.l.o.g. that $\psi(\varphi) = \psi(\varphi_i)$ for all i.
Let $\psi(\varphi) = t$. Now $C = \{\varphi_i, \varphi\}_{i \in \omega}$ is compact so $0_{C,\{t\}}$ is open in the compact-open topology and $\psi \in 0_{C,\{t\}}$. But then there is a j such that

$$i \geq j \to \psi_i \in 0_{C,\{t\}}$$

which means that

$$i \geq j \to \psi_i(\varphi_i) = t$$

so $\psi(\varphi) = \lim\limits_{i \to \infty} \psi_i(\varphi_i)$ and by Theorem 3.22 we see that $\psi = \lim\limits_{i \to \infty} \psi_i$ in T_{k+1}.

       □ Claim

We may now word by word repeat the proof of Lemma 3.43 establishing the equicontinuity of C.

       □

We are now ready to give a topological characterization of $\langle Ct(k), T_k \rangle$ due to M. Hyland. The result is mentioned without proof in [20].

Theorem 3.50

Let $k \geq 0$. Then T_{k+1} is the topology on $Ct(k+1)$ compactly generated from the compact-open topology on $Ct(k+1)$.

Proof: For $k = 0$ this result is trivial and left to the reader. For $k \geq 1$ we just have established

<u>i</u>   The compact-open topology on $Ct(k+1)$ is a subtopology of T_{k+1}.

<u>ii</u> T_{k+1} and the compact-open topology have the same compact sets.

<u>iii</u> T_{k+1} is compactly generated.

Together they give the theorem.

           □

Remark 3.51

The natural topology on the Cartesian product $Ct(k_1) \times \ldots \times Ct(k_n)$ inherited from the associates will also be compactly generated, and our characterizations of the compact subsets generalize easily. It follows that the compact sets will be the same as in the product topology. Thus the topology on $Ct(k_1) \times \ldots \times Ct(k_n)$ is the one compactly generated from the product-topology.

3.4   Filter-spaces and the countable functionals

In [20] Martin Hyland has given a thorough introduction to a filter-space discussion of the countable functionals, leading among other things to an abstract definition of recursion.

We will abstain from copying too much of Hyland [20]. But we think that the approach gives an interesting analysis of the structure of the countable functionals, and we will try to give an introduction to make the reader take an interest in the material. We urge the reader to go to the original paper [20] for further studies.

Continuity is a matter of respecting approximations. In a metric space or a topological space approximations by sequences or open neighbourhoods are the notions studied while in a general structure we may use even more abstract means to describe approximations. The idea of the notion of a filter-space is to capture some of the general properties of approximations.

Definition 3.52

<u>a</u>   Let X be a set. A <u>filter</u> F on X is a family of nonempty subsets Y of X such that

    <u>i</u> $Y_1 \subseteq Y_2 \wedge Y_1 \in F \rightarrow Y_2 \in F$

    <u>ii</u> $Y_1 \in F \wedge Y_2 \in F \rightarrow Y_1 \cap Y_2 \in F$

<u>b</u>   If G is a family of nonempty subsets of F such that any finite intersection from G is nonempty, we let the <u>filter generated from G</u> be

$$F[G] = \{Y \subseteq X ; \exists Y_1, \ldots, Y_k \in G(Y_1 \cap \ldots \cap Y_k \subseteq Y)\}$$

<u>c</u> $<X, F>$ is a <u>filter-space</u> if F maps every element $x \in X$ to a
family F_x of filters such that

<u>i</u> The principal ultrafilter $F[\{\{x\}\}]$ of x is in F_x .

<u>ii</u> If $F \in F_x$ and $F \subseteq G$, G a filter, then $G \in F_x$.

<u>d</u> Let $<X, F>$ and $<Y, G>$ be two filter-spaces, $\psi : X \to Y$. ψ is
said to be <u>continuous</u> if for all $x \in X$ and $F \in F_x$ we have
$F[G] \in G_y$ where $y = \psi(x)$ and $A \in G \leftrightarrow \exists B \in F \ A = \psi"B$.

Remark 3.53

The filters in F_x may be regarded as approximations to x .
Then ψ is continuous if it maps an approximation to x onto an
approximation to $\psi(x)$.

Now, if $<X, F>$, $<Y, G>$ are two filter-spaces, let Z be the
space of continuous functions from X to Y . We want to organize Z
to a filter-space.

Definition 3.54

Let $<X, F>$, $<Y, G>$ be filter-spaces, Z the class of continuous
functions from X to Y .

From each $f \in Z$, let $H \in H_f$ if the following holds:

Whenever $F \in F_x$ then $\{C \subseteq Y : \exists A \in F \ \exists B \in H \ (C = \bigcup\{g"A; g \in B\})\}$
generates an element in $G_{f(x)}$.

Let H be defined by $H(f) = H_f$. $<Z, H>$ is the function-filter-space.

Remark 3.55

It is a matter of routine to verify that $<Z, H>$ is a filter-space.

Now, let $k \geq 1$. We will organize each $Ct(k)$ to a filter-space.
If we forget our works on topology we see that the most natural way to
give approximations to a functional ψ is through the various B_σ^k's in
which ψ is an element. So we will try with the following definition.

Definition 3.56

Let $k \geq 1$.

<u>a</u> Let σ be such that $B_\sigma^{k-1} \neq \emptyset$ ($B_\sigma^0 = \{\sigma(0)\}$ by convention).
Let $V_{\sigma,t}^k = \{\psi \in Ct(k); \psi$ is constant t on $B_\sigma^{k-1}\}$.

<u>b</u> Let $\alpha \in As(k)$. Let F_α be the filter generated by

$$\{V_{\sigma,t}^k; \alpha(\sigma) = t+1\} .$$

<u>c</u>   Let $\Gamma \in F_\psi$ if there is an associate α for F such that $F_\alpha \subseteq \Gamma$.

<u>d</u> $Ct(0) = \omega$ is organized to a filterspace $<\omega, G>$ by letting G_n contain just the principal ultrafilter of n .

One part of the justification of this definition is given by

<u>Theorem 3.57</u>

Let $\Phi : Ct(k) \to \omega$. Then the following are equivalent.

<u>i</u> $\Phi \in Ct(k+1)$

<u>ii</u> Φ is continuous in the sense of the filter-space defined in 3.56.

<u>Proof</u>: If $k = 0$ this is easy, so we assume that $k \geq 1$.
Assume that $\Phi \in Ct(k+1)$, let $\psi \in Ct(k)$ and let $F \in F_\psi$. Choose an associate α for ψ such that $F_\alpha \subseteq F$. Let $\Phi(F)$ be the filter gene-rated from $\{\Phi"A ; A \in F\}$, $\Phi(F_\alpha)$ similarly defined. Since $F_\alpha \subseteq F$ we see that $\Phi(F_\alpha) \subseteq \Phi(F)$. So, in order to show continuity it is suffi-cient to show that $\Phi(F_\alpha)$ generates the principal ultrafilter of $\Phi(\psi)$.

Since $\Phi \in Ct(k+1)$ there is an n such that Φ is constant $\Phi(\psi)$ on $B_{\overline{\alpha}(n)}^k$. Let $\sigma_1, \ldots, \sigma_m$, $t_1, \ldots, t_m$ be those sequences and values such that each $\sigma_i < n$ and $\alpha(\sigma_i) = t_i + 1$.

Then $V = V_{\sigma_1 t_1}^k \cap \ldots \cap V_{\sigma_m t_m}^k$ is in F_α and $\Phi"V = \{\Phi(\psi)\}$, and $\{\Phi(\psi)\} \in \Phi(F_\alpha)$. This establishes <u>i</u> $\to$ <u>ii</u> .

Now assume that $\Phi : Ct(k) \to \omega$ is continuous. Let $\psi \in Ct(k)$, α an associate for ψ . It is sufficient to show that Φ is constant on $B_{\overline{\alpha}(n)}^k$ for some n .

By continuity of Φ , $\Phi(F_\alpha)$ generates the principal ultrafilter of $\Phi(\psi)$. This means that for some $V \in F_\alpha$ $\Phi"V = \{\Phi(\psi)\}$. W.l.o.g. we may assume that $V = V_{\sigma_1 t_1}^k \cap \ldots \cap V_{\sigma_m t_m}^k$ where $i \leq m \to \alpha(\sigma_i) = t_i + 1$.
Let $s = \max\{\sigma_i + 1 ; i \leq m\}$. Then $B_{\overline{\alpha}(s)}^k \subseteq V$ so Φ is constant on $B_{\overline{\alpha}(s)}^k$ But this was what we wanted.

□

The real justification comes through the next result which enables us to use filter-spaces to characterize the countable functionals. These results are due to Martin Hyland [20].

<u>Theorem 3.58</u>

Let $k \geq 1$ and let $Ct(k)$ and $Ct(0)$ be given their filter-space structures from Definition 3.56. Then the filter-space structures on $Ct(k+1)$ defined in 3.54 and 3.56 are the same.

Proof: We must establish two inequalities.

__i__ Let $\phi \in Ct(k+1)$ and let $H \in H_\phi$. We will show that there is an associate α for ϕ such that $F_\alpha \subseteq H$. This will show $H_\phi \subseteq F_\phi$.

Proof: We know from Definition 3.54 that for all $\psi \in Ct(k)$ and all associates β for ψ the following will hold:

$$\{C \subseteq \omega; \ \exists A \in F_\beta \ \exists B \in H \quad C = U\{g''A; g \in B\}\}$$

generates the principal ultrafilter of $\phi(\psi)$

so for some $A \in F_\beta$, $B \in H$ we have that all $g \in B$ are constant $\phi(\psi)$ on A .

In particular then there is a number n and a $B \in H$ such that all $g \in B$ are constant $\phi(\psi)$ on $B^k_{\bar{\beta}(n)}$ (see the end of the proof of Theorem 3.57).
Let

$$\alpha(\bar{\beta}(n)) = \begin{cases} t+1 & \text{if for some } B \in H \text{ all } g \in B \text{ are} \\ & \text{constant } t \text{ on } B^k_{\bar{\beta}(n)} \\ 0 & \text{otherwise} \end{cases}$$

By the considerations above, in order to show that α is an associate for ϕ , it is sufficient to show that whenever $\alpha(\bar{\beta}(n)) = t+1$ and β is an associate for ψ then $\phi(\psi) = t$. So let $\alpha(\bar{\beta}(n)) = t+1$. Let $B_1 \in H$ be such that all $g \in B_1$ are constant t on $B^k_{\bar{\beta}(n)}$.
We have established that there is an m and $B_2 \in H$ such that all $g \in B_2$ are constant $\phi(\psi)$ on $B^k_{\bar{\beta}(m)}$. But $B_1 \cap B_2 \neq \emptyset$ since H is a filter and $B^k_{\bar{\beta}(m)} \cap B^k_{\bar{\beta}(n)} \neq \emptyset$, so $t = \phi(\psi)$.
So α is an associate for ϕ . It remains to show that $F_\alpha \subseteq H$, and indeed it is sufficient to show that when $\alpha(\sigma) = t+1$ then $V^{k+1}_{\sigma,t} \in H$. Since $\alpha(\sigma) > 0$ there is a $B \in H$ such that all $g \in B$ are constant t on B^k_σ . Thus $B \subseteq V^{k+1}_{\sigma,t}$, and since $B \in H$ and H is a filter we see that $V^{k+1}_{\sigma,t} \in H$.

__ii__ Let $\phi \in Ct(k+1)$, α an associate for ϕ . We will show that $F_\alpha \in H_\phi$. This will show that $F_\phi \subseteq H_\phi$.

Proof: Let $\psi \in Ct(k)$, $F \in F_\psi$. We must show that

$$\{C \subseteq \omega; \ \exists A \in F \ \exists B \in F_\alpha \quad C = U\{g''A; g \in B\}\}$$

generates the principal ultrafilter of $\phi(\psi)$.

Let β be an associate for ψ . It is sufficient to show that $\exists A \in F_\beta \ \exists B \in F_\alpha$ (all $g \in B$ are constant $\phi(\psi)$ on A).
Choose n, t such that $\alpha(\bar{\beta}(n)) = t+1$ ($t = \phi(\psi)$). Let $B = V^{k+1}_{\bar{\beta}(n),t}$.

We let $A = V^k_{\sigma_1,t_1} \cap \ldots \cap V^k_{\sigma_m,t_m}$ where $\beta(\sigma_i) = t_i+1$ and $\sigma_1,\ldots,\sigma_m$ are those sequence-numbers $< n$ such that $\beta(\sigma_i) > 0$. Then $A \in F_\beta$ and $A = B^k_{\bar{\beta}(n)}$. But then, by definition

$$B = V^{k+1}_{\bar{\beta}(n),t} = \{g;\ g \text{ is constant } t \text{ on } A\}$$

and we are through.

□

From any filter-space $<X,F>$ we may derive a topology in the following way:

Definition 3.59

Let $<X,F>$ be a filter-space. Let the derived topology T_F be defined by

$$0 \in T_F \text{ if and only if } 0 \in F \text{ whenever } x \in 0 \text{ and } F \in F_x.$$

Remark 3.60

<u>a</u>   It follows easily from the properties of a filter that T_F is a topology.

<u>b</u>   We see that the more filters present in each F_x, the weaker will T_F be.

<u>c</u>   It is impossible to recover F from T_F in general, since the filter-space generated from the open neighbourhood filters in T_F may be properly included in F. We will see an example of this in our next result.

Theorem 3.61

Let $<Ct(k),F>$ be the filter-space defined in 3.56. Then $T_k = T_F$.

Proof:

<u>i</u>   Let $0 \in T_F$, $\psi \in 0$. Let α be an associate for ψ. Then $F_\alpha \in F_\psi$ so $0 \in F_\alpha$. Then there are $\sigma_1,\ldots,\sigma_m$, $t_1,\ldots,t_m$ such that $\alpha(\sigma_i) = t_i+1$ $(i = 1,\ldots,m)$ and $V^k_{\sigma_1,t_1} \cap \ldots \cap V^k_{\sigma_m,t_m} \subseteq 0$. Choose $n > \max\{\sigma_1,\ldots,\sigma_m\}$. Then $B^k_{\bar{\alpha}(n)} \subseteq V^k_{\sigma_1,t_1} \cap \ldots \cap V^k_{\sigma_m,t_m} \subseteq 0$. But this shows that $0 \in T_k$.

<u>ii</u>   Now assume that $0 \in T_k$. Let $\psi \in 0$ and let $F \in F_\psi$. We must show that $0 \in F$.
By definition of F_ψ there is an associate α for ψ such that $F_\alpha \subseteq F$, and it is sufficient to show that $0 \in F_\alpha$. Pick m such that

$B_{\bar{\alpha}(m)}^{k} \subseteq 0$. We saw in the proof of Theorem 3.57 an argument showing that $B_{\bar{\alpha}(m)}^{k} \in F_{\alpha}$. But then $0 \in F_{\alpha}$ since all supersets of $B_{\bar{\alpha}(m)}^{k}$ will be in F_{α} .

□

We will now conclude this section by giving a set of closed neighbourhoods U , which may serve as an alternative to the B_{σ}^{k}'s .

Definition 3.62

<u>a</u>   Let $U^{0} = \omega$.

Inductively, let $U^{k+1} = \{<<u_1,t_1>,\ldots,<u_n,t_n>>; u_1,\ldots,u_n \in U^k, t_1,\ldots,t_n \in \omega\}$

<u>b</u>   If $u \in U^k$ we define $U_u \subseteq Ct(k)$ by induction as follows

If $u \in U^0$, let $U_u = \{u\}$.

If $u = <<u_1,t_1>,\ldots,<u_n,t_n>> \in U^{k+1}$, let

$$U_u = \{\psi \in Ct(k+1) ; \psi \text{ is constant } t_1 \text{ on } U_{u_1} \wedge \ldots \wedge \psi \text{ is constant } t_n \text{ on } U_{u_n} \}$$

Remark 3.63

The formal neighbourhood $u \in U^k$ is independent of the type-structure we are working with, while the interpretation U_u depends on $Ct(k)$. The structure of the u's give a skeleton of the various structures on $<Ct(k)>_{k \in \omega}$ and on related type-structures. Using methods of chapter 5 we will see that the relation $U_{u_1} \cap U_{u_2} \neq \emptyset$ is primitive recursive, and the same relation holds for all type-structures closed under computations. Most of the theory we have discussed in this chapter could have been based on $<U^k>_{k \in \omega}$ instead of on the associates. We would just have to find some alternative definition of $Ct(k)$ to start with.

As a general conclusion to this chapter we will claim that the characterizations of the countable functionals we have given show that $<Ct(k)>_{k \in \omega}$ is a natural hierarchy of functionals obtained by iteration of standard constructions of function-spaces. The fact that these constructions fit together adds to this conclusion.

4. COMPUTABILITY VS RECURSION

4.1 Degrees of functionals

In chapter 2 we defined the countable functionals, Kleene-computations and recursions. Our investigations of these notions were still at a rather elementary level.

In this chapter we will investigate and compare the degrees of unsolvability derived from these two notions. There are, however, several problems concerning the ordertype of these degrees that we leave unsolved and untouched. See also Gandy-Hyland [13] for a general discussion.

From now on and throughout the book all notions will unless otherwise is stated, refer to the countable type-structure.

We have selected some problems which we think are useful when we want to compare our two notions for algorithms on the functionals.

Later, in chapters 5 and 6, when we develop some machinery in order to investigate envelopes and sections, we will also get some feed-back to properties of the degrees.

The degree-structure of recursions is coarser than the degree-structure of computations. This is a direct consequence of Theorem 2.28. We will see that it is strictly coarser.

Definition 4.1

a   Let ψ, φ be two countable functionals.
ψ is (Kleene-) _equivalent_ to φ if ψ is computable in φ and φ is computable in ψ .

b   If ψ is a countable functional, let the _degree of ψ_ , dg ψ be the equivalence-class of ψ under the relation of a .
We impose a partial ordering on the degrees by

$$\text{dg } \psi \leq \text{dg } \varphi$$

if ψ is computable in φ .

c   Let $\underset{\sim}{d}$ be a degree. By $\text{Tp}(\underset{\sim}{d})$ = the type of $\underset{\sim}{d}$ we mean the least number k such that $\underset{\sim}{d} \cap \text{Ct}(k) \neq \emptyset$, i.e. the least k such that $\exists \varphi \in \text{Ct}(k) \quad \underset{\sim}{d} = \text{dg } \varphi$.

d   Let $\psi \in \text{Ct}(k)$. We call ψ _irreducible_ if $\text{Tp}(\text{dg } \psi) = k$ i.e. if ψ is not equivalent to any functional of type $< k$.

e   Let $\underset{\sim}{d}$ be a degree. By the _cone of $\underset{\sim}{d}$_ we mean $\cup \{\underset{\sim}{e}; \underset{\sim}{d} \leq \underset{\sim}{e}\}$.

<u>f</u>   A countable functional ψ is k-obtainable if ψ is computable in some functional of type k. $\psi \in Ct(k)$ is <u>non-obtainable</u> if ψ is not k'-obtainable for any $k' < k$.

<u>g</u>   A degree $\underset{\sim}{d}$ is <u>minimal</u> if $\underset{\sim}{d} \neq \underset{\sim}{0}$ (the degree of the computable functionals) but

$$\forall \underset{\sim}{e} < \underset{\sim}{d} \; (\underset{\sim}{e} = \underset{\sim}{0})$$

<u>Remark 4.2</u>

<u>a</u>   If $\psi \in Ct(k)$ then ψ is k'-obtainable for all $k' \geq k$. (Lemma 4.3)

<u>b</u>   A functional $\psi \in Ct(k)$ is non-obtainable if k is the minimal type occuring in the cone of ψ.

In this section we will prove some elementary results about the degrees and discuss some open problems.

<u>Lemma 4.3</u>

Let $\underset{\sim}{d}$ be a degree, $k = Tp(\underset{\sim}{d})$. For all $k' \geq k$ there are elements of type k' in $\underset{\sim}{d}$.

<u>Proof</u>: Let $P_k^{k'}$ and $P_{k'}^{k}$ be the push up and push down operators from Definition 1.7. These are computable. Let $\psi \in \underset{\sim}{d} \cap Ct(k)$, $\Phi = P_k^{k'}(\psi)$. Then $\psi = P_{k'}^{k}(\Phi)$, so $dg\,\psi = dg\,\Phi = \underset{\sim}{d}$.

▫

<u>Lemma 4.4</u>

Let $\{\underset{\sim}{d}_i\}_{i \in \omega}$ be a sequence of degrees of type k. Then this sequence is bounded by a degree of type k.

<u>Proof</u>: Let $\{\psi_i\}_{i \in \omega}$ be a sequence from $Ct(k)$ such that for all $i \in \omega$ we have $\psi_i \in \underset{\sim}{d}_i$.
Define $\psi \in Ct(k)$ by

$$\psi(<\varphi_0, \varphi_1>) = \psi_{P_{k-1}^0(\varphi_0)}(\varphi_1)$$

Each ψ_i is computable in ψ by

$$\psi_i(\varphi) = \psi(<P_0^{k-1}(i), \varphi>)$$

so $\underset{\sim}{d}_i \leq dg\,\psi$.

▫

By Lemma 4.4 all sequences of degrees from $Ct(k)$ are bounded in the natural ordering. It is then natural to ask whether all sequences

of degrees are bounded in this ordering. Combining Lemma 4.5 with
Corollary 7.19 we see that this is not the case.

Lemma 4.5

The following statements are equivalent.

i   There is a number k such that all functionals are k-obtainable.

ii   All sequences $\{\underset{\sim}{d}_i\}_{i \in \omega}$ of degrees are bounded.

Proof:

i $\Rightarrow$ ii   Assume i and let $\{\underset{\sim}{d}_i\}_{i \in \omega}$ be a sequence of degrees. By i and
the proof of Lemma 4.4 we see that this sequence is bounded by a
degree of type k .

ii $\Rightarrow$ i   Assume ii. Let ψ_k be a functional that is not k-obtainable
and let $\underset{\sim}{d}_k = dg \, \psi_k$. If $\{\underset{\sim}{d}_k\}_{k \in \omega}$ is bounded by a degree $\underset{\sim}{d}$,
let $\phi \in \underset{\sim}{d}$ and let k_0 be the type of ϕ .
For $k \geq k_0$ we get ψ_k computable in ϕ . But ψ_k is not k-obtainable,
and by Lemma 4.3, ψ_k is not k_0-obtainable. This is a contradiction.

$\square$

Remark 4.6

Later we will construct countable nonobtainable functionals of
any type ≥ 3 . As we have seen in Theorem 2.23 all functionals in Ct(2)
are 1-obtainable.

A nonobtainable functional is clearly irreducible. So, when we
have shown that there are nonobtainable functionals, we have in particular shown that some degrees are of type > 1 . In the next section we
will show that there are irreducible functionals of type 2 .

Remark 4.7

From ordinary recursion theory it is known that there is a minimal
Turing degree. After the manuscript was typed it has been shown by the
author that there is no minimal Kleene-degree over $<Ct(k)>_{k \in \omega}$.

The type 2 continuous functionals are stratified into a hierarchy
by the Kalmar rank defined as follows.

Definition 4.8

Let $F \in Ct(2)$, $T_F = \{\sigma; F \text{ is not constant on } B_\sigma\}$.
(B_σ is defined in 2.16.)

T_F is a well-founded tree, and the Kalmar rank of F is the
height of that tree.

Remark 4.9

The Kalmar rank KR may be defined as follows:
If F is constant, then $KR(F) = 0$. Otherwise let $F_n(\alpha) = F(n^{\wedge}\alpha)$
where

$$(n^{\wedge}\alpha)(0) = n , \quad (n^{\wedge}\alpha)(k+1) = \alpha(k)$$

Then $KR(F) = \text{Sup}\{KR(F_n)+1; n \in \omega\}$.

The irreducible functionals we construct will be of Kalmar rank ω
If $\underset{\sim}{d}$ is a degree of type ≤ 2, we let $KR(\underset{\sim}{d})$ be the minimal Kalmar
rank of the elements of $\underset{\sim}{d} \cap Ct(2)$.
The following problem was communicated to me by Jan Bergstra.

Problem 4.10

Are there degrees of type 2 of arbitrary high Kalmar rank ?

4.2 Irreducible functionals of type 2

In this section we will go a bit deeper into the degree-theory of
the continuous type-2 functionals. We will start with a construction
of an irreducible functional. This construction is due to J. Bergstra
[1], based on an idea of Martin Hyland. Then we will construct a non-
computable functional F such that for all $H \in Tp(2)$

$$1\text{-sc} (F,H) = 1\text{-sc} (H)$$

These functionals will all be computable when restricted to the
recursive functions. This will be of importance to our arguments. In
the end of this section we will give an unpublished construction due to
Leo Harrington, where he produces an F which is not computable, even
when restricted to the recursive functions, but still with only recursive functions in the 1-section.

First we need a simple but important result.

Theorem 4.11

Let $F,H \in Tp(2)$ and assume that there is a partial computable
functional $G \subseteq F$ such that G is defined on all H-computable functions.
Then $1\text{-sc}(F,H) = 1\text{-sc}(H)$.

Proof: $\supseteq$ is obvious, so we must establish $\subseteq$. The main point of the argument is that whenever we in a computation in F,H want to refer to F , we may instead refer to the computable G . This point is precisely formulated as follows.

Claim   Assume that $\{e\}(F,H,n_1,\ldots,n_k) \simeq m$.

Then $\{e\}(G,H,n_1,\ldots,n_k) \simeq m$

The claim is proved by induction on the length of the computation $\{e\}(F,H,n_1,\ldots,n_k)$.

The only case where there is anything to prove at all, is when S8 is used at F , so assume

$$\{e\}(F,H,n_1,\ldots,n_k) = F(\lambda n\{e_1\}(F,H,n,n_1,\ldots,n_k))$$

Let $\alpha = \lambda n\{e_1\}(F,H,n,n_1,\ldots,n_k)$. By the induction hypothesis

$$\alpha = \lambda n\{e_1\}(G,H,n,n_1,\ldots,n_k)$$

Since G is computable we see that α is computable in H , so G is defined at α and $G(\alpha) = F(\alpha)$. But then

$$\{e\}(G,H,n_1,\ldots,n_k) = G(\alpha) = F(\alpha) = \{e\}(F,H,n_1,\ldots,n_k)$$

This establishes the claim.

The theorem now follows, because if $\alpha = \lambda n\{e\}(F,H,n)$ then by the claim $\alpha = \lambda n\{e\}(G,H,n)$ so α is computable in H .

□

We are now going to construct an irreducible functional. This was first done by Peter Hinman [17] using a so-called spoiling argument. It is also used in Bergstra [1]. We will not go into this method here. Instead we will use an alternative method from Bergstra [1].

Definition 4.12

Let $T(e,x,y)$ be Kleene's T-predicate with the following proper-ties:

i   T is primitive recursive

ii   Let $W_e = \{x; \exists y\ T(e,x,y)\}$. Then all r.e. sets are of the form W_e for some e .

iii   Given e,x there is at most one y such that $T(e,x,y)$. Moreover

$$\forall y\ (T(e,x,y) \rightarrow y > 0)$$

Remark 4.13

Kleene's T-predicate or some equivalent predicate is defined in almost any textbook in elementary recursion theory, e.g. Rogers [38].

Definition 4.14

<u>a</u>   Let $f: \omega \to \omega$. f is a <u>modulus for W_e</u>   if for all x

$$x \in W_e \leftrightarrow \exists y \, T(e,x,y) \leftrightarrow \exists y \leq f(x) \, T(e,x,y)$$

In many texts f will be called a computation function for W_e .

<u>b</u>   If τ is a sequence number we say $\tau \in \text{Mod}(e)$ or   <u>$\text{Mod}(e,\tau)$</u>   if

$$\forall x < \text{lh}(\tau) \, (\exists y < \text{lh}(\tau) T(e,x,y) \to \exists y \leq \tau(x) T(e,x,y))$$

<u>c</u>   Given e, e' , let

$$F_e^{e'}(x,f) = \begin{cases} 0 & \text{if } \exists n \, \text{Mod}(e, \bar{f}(n)) \wedge T(e',x,n) \\ 1 & \text{otherwise} \end{cases}$$

Remark 4.15

<u>a</u>   If f is a modulus for W_e then $f(x)$ gives an upper bound for the y's we must check through in order to decide if $x \in W_e$.   If we compare this definition with Definition 3.28   we see that f will be a modulus-function for the sequence

$$W_e^n = \{x; \exists y \leq n \, T(e,x,y)\} \quad (n = 0,1,\dots)$$

<u>b</u> $\text{Mod}(e,\tau)$ means that if we want to find an example showing that τ cannot be extended to a modulus for W_e , then we have to regard W_e^n's for $n \geq \text{lh}(\tau)$, so τ "looks like" a beginning of a modulus for W_e .

Lemma 4.16

<u>a</u>   Mod   is recursive, and if τ is a subsequence of σ and $\neg \text{Mod}(e,\tau)$ then $\neg \text{Mod}(e,\sigma)$.

<u>b</u>   If W_e is not recursive in f , then there is an n such that $\neg \text{Mod}(e, \bar{f}(n))$.

<u>c</u> $F_e^{e'}$ is computable in $W_{e'}$, and thus continuous.

<u>d</u> $W_{e'}$ is computable in W_e , $F_e^{e'}$.

<u>e</u>   There is a partial computable function $G_e^{e'} \subseteq F_e^{e'}$ such that $G_e^{e'}(x,f)$ is defined whenever W_e is not recursive in f .

Proof:

<u>a</u>   is best proved by inspection.

<u>b</u>   We give a contrapositive argument. Assume $\forall n \, \text{Mod}(e, \bar{f}(n))$. Then f is a modulus for W_e, and W_e is recursive in f by

$$x \in W_e \longleftrightarrow \exists y \leq f(x) T(e,x,y)$$

<u>c</u>   We give the following algorithm to compute $F_e^{e'}(x,f)$ from $W_{e'}$:

      If $x \notin W_{e'}$, let $F_e^{e'}(x,f) = 1$

      If $x \in W_{e'}$, let n be such that $T(e',x,n)$

If $\text{Mod}(e, \bar{f}(n))$, let $F_e^{e'}(x,f) = 0$, otherwise let $F_e^{e'}(x,f) = 1$.

<u>d</u>   Let W_e be given. Let

$$f(x) = \begin{cases} \mu n \, T(e,x,n) & \text{if } x \in W_e \\ 0 & \text{if } x \notin W_e \end{cases}$$

Then f is recursive in W_e and f is a modulus for W_e, so for all n

$$\text{Mod}(e, \bar{f}(n))$$

But then

$$F_e^{e'}(x,f) = 0 \quad \text{iff} \quad \exists n \, T(e',x,n) \quad \text{iff} \quad n \in W_{e'}$$

This shows that $W_{e'}$ is computable in W_e, $F_e^{e'}$.

<u>e</u>   Define $G_e^{e'}$ by the following partial algorithm:

    Let y be minimal such that $\neg \text{Mod}(e, \bar{f}(y))$.

(If no such y exists, $G_e^{e'}(x,f)$ will be undefined, but then f is a modulus for W_e and W_e is recursive in f.)

Then let

$$G_e^{e'}(x,f) = \begin{cases} 0 & \text{if } \exists n < y \, T(e',x,n) \\ 1 & \text{if } \forall n < y \, \neg T(e',x,n) \end{cases}$$

$G_e^{e'}$ is computable since we have given an algorithm for it. It is easy to see that if $G_e^{e'}(x,f)$ is defined, then $G_e^{e'}(x,f) = F_e^{e'}(x,f)$, and if $G_e^{e'}(x,f)$ is undefined, then W_e is recursive in f.   □

We are now ready to show

Theorem 4.17

There exists a continuous irreducible functional of type 2.

Proof: Choose e, e' such that W_e is not recursive, $W_{e'}$ is not recur sive in W_e . (This is possible by the Friedberg-Mucnik solution of Post's problem (see e.g. Rogers [38], Shoenfield [43] or Sacks [39]).) We will show that $F_e^{e'}$ is irreducible.

W_e is not recursive, so by Lemma 4.16 $\underline{e}$, $G_e^{e'}(x,f)$ is defined whenever f is recursive. If we let $H = {}^2 0$ in Theorem 4.11 we see that

$$1-sc(F_e^{e'}) = 1-sc(F_e^{e'},{}^2 0) = 1-sc({}^2 0)$$

which contains only recursive functions.
So, if $F_e^{e'}$ is equivalent to a function f , f must be recursive, so $F_e^{e'}$ is computable. But then

$$1-sc(F_e^{e'},W_e) = 1-sc(W_e)$$

contradicting the fact that $W_{e'}$ is computable in $F_e^{e'}$, W_e (Lemma 4.16 $\underline{d}$) while $W_{e'}$ is not recursive in W_e .
It follows that $F_e^{e'}$ is irreducible.

□

In this argument we used the facts that W_e is not recursive and that $W_{e'}$ is not recursive in W_e in order to show that $F_e^{e'}$ is irreducible. Now, if W_e is not recursive, will then F_e^e be irreducible ? We can not answer this problem in general, not knowing the correct answer. But, if F_e^e is irreducible we will have a non-computable continuous functional of type 2 with the following remarkable property.

Lemma 4.18

$$\forall H \in Tp(2)[1-sc(F_e^e,H) = 1-sc(H)]$$

Proof: We regard two cases.

$\underline{i}$ W_e is computable in H .
Then F_e^e is computable in H , by Lemma 4.16 $\underline{c}$, so

$$1-sc(F_e^e,H) = 1-sc(H)$$

$\underline{ii}$ W_e is not computable in H . Then, by Lemma 4.16 $\underline{e}$, G_e^e is total on $1-sc(H)$, and by Theorem 4.11

$$1-sc(F_e^e,H) = 1-sc(H)$$

□

J. Bergstra [1] constructed two continuous functionals F and G such that for all $H \in Tp(2)$

$$1-sc(F,H) = 1-sc(G,H)$$

but F and G are not recursively equivalent. This shows that the family $\{1-sc(F,H); H \in Tp(2)\}$ is not sufficient to characterize F up to equivalence. As we have mentioned, if we could find an irreducible F_e^e, we would have a slight improvement of Bergstra's result. We will not do this directly, but construct an irreducible functional having all the properties of an F_e^e, so Lemma 4.18 will work.

We will now fix some notation for the next lemma.

Let $T \subseteq \omega \times \omega$ be recursive, $S_T = \{e; \exists s T(e,s)\}$. A <u>modulus for S_T</u> is a function f such that

$$\forall e(e \in S_T \leftrightarrow \exists s \leq f(e) T(e,s))$$

$$Mod_T(\tau) \leftrightarrow \forall e < lh(T) \, (\exists s < lh(\tau) T(e,s) \rightarrow \exists s \leq \tau(e) T(e,s))$$

$$F_T(e,f) = \begin{cases} 0 & \text{if } \exists s \, Mod_T(\bar{f}(s)) \wedge T(e,s) \\ 1 & \text{otherwise} \end{cases}$$

There is nothing in the arguments for F_e^e we have given that would not hold for F_T, so if we can construct T such that F_T is irreducible, then we will have an improvement of Bergstra's result.

<u>Lemma 4.19</u>

We use the notation described above.
There is a recursive T such that F_T is not computable.

<u>Proof</u>: By induction on $s \in \omega$ we will define the finite set $T_s \subseteq \omega \times \omega$, and we will let $T = \{(e,s); (e,s) \in T_{s+1}\}$. We will also let

$$\alpha_s(e) = \begin{cases} \text{least } t \text{ such that } (e,t) \in T_s & \text{if such } t \text{ exists} \\ 0 & \text{otherwise} \end{cases}$$

α_s will be a modulus for T_s.

<u>Construction</u> Let $T_0 = \emptyset$.
Let $s = \langle e,n \rangle$ (by some standard surjective coding of ordered pairs) and assume that T_s is constructed, with α_s defined as above. If there is an $s' < s$ such that $(e,s') \in T_s$, let $T_{s+1} = T_s$. Assume that there is no such s'.
If $\{e\}_s(e,\alpha_s) = 1$, let $T_{s+1} = T_s \cup \{(e,s)\}$. Otherwise let $T_{s+1} = T_s$. ($\{e\}_s(e,\alpha_s) = 1$ means that $\{e\}(e,\alpha_s) = 1$ by a computation of $\leq s$ steps.)

It is easy to see that T defined as above is recursive.

<u>Claim</u> For all s we have $Mod_T(\bar{\alpha}_s(s))$.

Proof: We must show that for all $e < s$ we have that

$$\exists n < s\, T(e,n) \leftrightarrow \exists n \leq \alpha_s(e) T(e,n)$$

$\leftarrow$: By a trivial induction on $s > 0$ we see that $e < s \to \alpha_s(e) < s$, so $\leftarrow$ is easy.

$\to$: If for some $n < s$ we have $T(e,n)$, we must have that $(e,n) \in T_s$ since for other elements of T the second coordinate will be $\geq s$. But then $\alpha_s(e) = n$ by the definition of α_s . This establishes $\to$.

Now, to show that F_T is not computable, we assume the converse, so assume that e_0 is such that

$$F_T(e,f) = \{e_0\}(e,f)$$

for all e,f . We will regard two cases.

$\underline{i}$ There is an s such that $T(e_0,s)$.

Then, by the construction of T , $(e_0,s) \in T_{s+1}$ and $\{e_0\}_s(e_0,\alpha_s) = 1$. By the claim we have $\text{Mod}(\bar{\alpha}_s(s))$ so $F_T(e_0,\alpha_s) = 0$ since $\text{Mod}(\bar{\alpha}_s(s)) \wedge T(e_0,s)$ while $F_T(e_0,\alpha_s) = \{e_0\}(e_0,\alpha_s) = 1$, a contradiction

$\underline{ii}$ There is no s such that $T(e_0,s)$.

Let α be the canonical modulus for T , i.e.

$$\alpha(e) = \begin{cases} \mu s\, T(e,s) & \text{if } \exists s\, T(e,s) \\ 0 & \text{otherwise} \end{cases}$$

Then $F_T(e_0,\alpha) = 1$, so $\{e_0\}(e_0,\alpha) = 1$.

Now $\alpha = \lim\limits_{s \to \infty} \alpha_s$. Let s_0 be such that for all $s \geq s_0$ we have that $\{e_0\}_s(e_0,\bar{\alpha}(s)) = 1$, and let $s_1 > s_0$ be such that $\forall s \geq s_1 (\bar{\alpha}(s_0) = \bar{\alpha}_s(s_0))$.

Let $s \geq s_1$ be of the form $<e_0,n>$. By the definition of T_{s+1} we would have $(e_0,s) \in T_{s+1} \subseteq T$. But this contradicts the assumption. It follows that F_T cannot be computable.

□

Theorem 4.20

There is a continuous irreducible functional F of type 2 such that

$$\forall H \in \text{Tp}(2)\ (1\text{-sc}(H) = 1\text{-sc}(H,F))$$

Proof: We let $F = F_T$ as constructed in Lemma 4.19. Then F is irreducible and continuous. The argument of Lemma 4.18 is valid for F_T as well as for F_e^e .

□

A common property of all the functionals we have considered in this section is that if we restrict them to the recursive functions, then they become computable. One might ask if this reflects a general property of continuous functionals of type 2 , i.e. if 1-sc(F) consists of recursive functions only when there is a partial computable subfunction of F defined on all recursive functions ?

Leo Harrington constructed a counterexample to this by showing

Theorem 4.21

There exists a continuous functional F of type 2 such that

i If f is recursive in F , then f is recursive

ii For all e , if $\lambda f\{e\}(f) \subseteq F$, then there is a recursive f such that $\{e\}(f)$ is undefined.

Proof: The proof is by a priority-argument.

By induction on s we will define associates α_s approximating an associate α for F . Recall the definition $B_\sigma = \{f; \bar{f}(lh(\sigma)) = \sigma\}$.

We want to satisfy the following set of conditions:

C_n : $F \restriction B_{<n>}$ is continuous

I_e : If $\lambda n [e](n,F)$ is total, then $\lambda n[e](n,F)$ is recursive

N_e : $F \neq \lambda f\{e\}(f)$

(Recall the definition of the recursion

$$[e](n,F) \simeq k \leftrightarrow \forall \alpha \in As(F) \; \{e\}(n,\alpha) \simeq k)$$

At each stage of the construction, let F_s be the functional with associate α_s .

We will satisfy C_n by making sure that there will be an s such that $F \restriction B_{<n>} = F_s \restriction B_{<n>}$.

In order to satisfy I_e we want to protect $[e](n,F)$ for all n . One problem is that we cannot effectively decide if $[e](n,F_s)\downarrow$. We could solve this by just regarding $\{e\}(n,\alpha_s)$. But if we were going to protect such computations for all n (i.e. not permitting to change the part of α_s entering into such a computation) we would probably fix α in a recursive way. So, we will regard other associates for F_s , namely

$$\alpha_s^n(\tau) = \begin{cases} \alpha_s(\tau) & \text{if } lh(\tau) \geq n \\ 0 & \text{otherwise} \end{cases}$$

and in order to satisfy I_e we will try to protect computations of the form $\{e\}(n,\alpha_s^n)$.

We see that when we want to satisfy C_n or I_e , we find a part of the domain of F and say that we want F to be equal to F_s on this part.

In order to satisfy N_e we will look for a τ such that $\{e\}(\tau) = k$ for some k , and we want F to be different from k on B_τ . This may clearly lead to a conflict if we want F to be constant k on B_τ in order to satisfy some C_n or $I_{e'}$.

Such conflicts are solved by assigning priority to the conditions, we give a listing $\{R_i\}_{i \in \omega}$ of all the conditions and say that if $i < j$, then R_i has <u>higher priority</u> than R_j . This means that if $i < j$, then we may act in conflict with the interest of R_j in order to satisfy R_i . The listing will be effective, so that we may recursively in i find the condition R_i .

The construction is by stages s . At each stage s we will regard one of the conditions such that we may recursively decide which condition to regard, and each condition will be regarded cofinally often during the construction.

In order to satisfy R_i at stage s we may create a requirement B for R_i at stage s . B will be a finite union of B_σ's . B is <u>active</u> at a stage $s' > s$ if $F_s \upharpoonright B = F_{s+1} \upharpoonright B$, otherwise B is inactive. In order to satisfy C_n or I_e we will just create requirements. In order to satisfy N_e at stage s we may want $F_{s+1} \neq F_s$. But, if f is in an active requirement of some condition of higher priority, we must have $F_{s+1}(f) = F_s(f)$.

We will now give the precise construction.

Let $\alpha_0(\tau) = 1$ for all $\tau \neq < >$, $\alpha_0(< >) = 0$. Thus F_0 is the constant zero functional.

Assume that α_s is constructed and that at stage s we regard R_i .

<u>Case 1</u>: $R_i = C_n$ Let $\alpha_{s+1} = \alpha_s$. Let $B_{<n>}$ be a requirement for R_i . Proceed to stage $s+1$.

<u>Case 2</u>: $R_i = I_e$ Let $\alpha_{s+1} = \alpha_s$.
Look for a minimal $n \leq s$ such that

<u>i</u> $\forall m \leq n \; \{e\}_s(m, \alpha_s^m) \downarrow$

<u>ii</u> $\forall m < n$ there is an active requirement for e, m at stage s .

<u>iii</u> There is no active requirement for e, n at stage s .

(By some book-keeping device we may assume that we at each stage know

the requirements created, if they are active and for what purpose they
are created.)

If there is no such $n \leq s$ proceed to stage $s+1$.

If there is such an n , let $\tau_1,\ldots,\tau_k$ be the sequences such that
$\alpha_s^n(\tau_i) > 0$ and $\alpha_s^n(\tau_i)$ is used in the computation of $\{e\}_s(n,\alpha_s^n)$.

Let $B = B_{\tau_1} \cup \ldots \cup B_{\tau_k}$ be a requirement for e,n .

Proceed to stage $s+1$.

Case 3: $\underline{R_i = N_e}$ If there is a requirement for R_i active at stage s,
   let $\alpha_{s+1} = \alpha_s$ and proceed to stage $s+1$.

If there is no such requirement for R_i , look for a sequence-number
$\tau \leq s$ such that

i B_τ is not a subset of any active requirement for a condition of
   higher priority than R_i .

ii $\{e\}_s(\tau)\downarrow$ (here we treat τ as the beginning of an f , i.e. as
   a partial function).

If there is no such τ , let $\alpha_{s+1} = \alpha_s$ and proceed to stage $s+1$.
Otherwise, pick the least such τ , let τ_0 be the least extension of τ
such that $B_{\tau_0} \cap B = \emptyset$ for all active requirements of conditions of
higher priority than R_i (such τ_0 will exist) and let

$$
\alpha_{s+1}(\sigma) = \begin{cases} \{e\}_s(\tau)+2 & \text{if } \tau_0 \preceq \sigma \\ 0 & \text{if } \sigma \prec \tau_0 \\ \alpha_s(\sigma) & \text{if } \sigma \text{ and } \tau_0 \text{ are incomparable} \end{cases}
$$

Let B_{τ_0} be a requirement for N_e and proceed to stage $s+1$.

This ends the construction.

Claim

a   Each α_s is uniformly recursive in s .

b   For each i there is a stage s such that any requirement for R_i
   created after stage s remains active throughout the construction.
   If R_i is of the form C_n then $\alpha_{s'}(\tau) = \alpha_s(\tau)$ for all τ such
   that $\tau(0) = n$ and all $s' \geq s$.

Proof:

a   The construction is indeed by primitive recursion in s . At each
   stage there is a finite set of requirements constructed and the
   nature of these requirements is easily describable in a primitive

recursive way.

<u>b</u>   The only way a requirement from R_i can turn from active to in-
active is if we try to satisfy some I_e of higher priority than
R_i . If we create a requirement for I_e , we will not try to satisfy
it again unless this requirement becomes inactive. It follows by an
easy induction on i that we cannot destroy requirements for R_i more
than finitely many times. (The number of stages at which we may injure
a requirement for R_i is bounded by the function

$$h(0) = 0$$

$$h(n+1) = 2h(n)+1 \quad)$$

The second part of <u>b</u> is seen as follows. Let R_i be C_n and let s_0
be as in the first part of <u>b</u> . Let $s > s_0$ be such that we regard R_i
at stage s . The requirement we create at stage s will never be inactive, which means that if $s' \geq s$ and $\tau(0) = n$, then
$\alpha_{s'+1}(\tau) = \alpha_{s'}(\tau) = \alpha_s(\tau)$.

A consequence of Claim 1 <u>b</u> is that the sequence α_s has a limit α
which is an associate. We let F be the functional which has α as
an associate.
We will show that F satisfies the theorem.

First assume that f is recursive in F . Then there is an index
e such that for all n

$$f(n) = [e](n,F)$$

We will show that f is recursive.
Choose s such that after stage s no active requirement for I_e
is ever made inactive. This means that if we want to freeze a computation

$$\{e\}(n,\alpha_{s'}^n)$$

as some stage $s' \geq s$, we will succeed.
Moreover, since $\lambda n[e](n,F)$ is total we have that

$$\{e\}(n,\alpha^n)\!\downarrow$$

for all n .
Since $\alpha_{s'}^n \to \alpha^n$ there will be some $s' \geq s$ such that
$\{e\}(n,\alpha^n) = \{e\}_{s'}(n,\alpha_{s'}^n)$.
So, when we try to satisfy I_e we will for each n create a requirement for e,n at some stage $s' \geq s$, and this requirement will never
be made inactive.

So the following will be an algorithm for computing $f = \lambda n [e](n,F)$:

Given n go to the least $s' \geq s$ such that at stage s' there is an active requirement for e,n and then

$$f(n) = \{e\}_{s'}(n, \alpha_{s'}^n)$$

It remains to show that if $\lambda f\{e\}(f) \subseteq F$, then $\{e\}(f)$ is undefined for some recursive f.

If there is a requirement B_τ for N_e that remains active throughout the construction, F will be constant $k+1$ on B_τ where $k = \{e\}(f)$ for any $f \in B_\tau$. To see this, go back to case 3: $R_i = N_e$. Then F and $\lambda f\{e\}(f)$ takes conflicting values on B_τ.

So, if $\lambda f\{e\}(f) \subseteq F$, N_e will have no permanently active requirements, and by Claim 1 there will be a stage s after which no requirements for N_e are created.

<u>Claim 2</u> There is a recursive f such that f is in no requirement for any condition of higher priority than N_e.

<u>Proof</u>: Let s be a stage satisfying: If for some R_i of higher priority than N_e only finitely many requirements are created, these are all created before stage s. Let $B_{\tau_1} U \ldots U B_{\tau_m}$ be the union of these requirements.

Let $e_1, \ldots, e_k$ be those indices such that I_{e_i} has higher priority than N_e and we create infinitely many requirements for I_{e_i}. Define f by

$$f(0) = 1 + Max\{\tau_i(0); i \leq m\}$$

For $n > 0$ let

$f(n) = t+1$ where t is maximal such that there is an $i \leq k$ and a requirement $B_{\sigma_1} U \ldots U B_{\sigma_m}$ for $e_i, n+1$ active at a stage $> s$ and a $j \leq m$ such that $t = \sigma_j(n)$.

f is recursive since, given n, we may compute all actual $\sigma_1, \ldots, \sigma_m$ and then find t.

By the choice of $f(0)$, f is in no requirement for a condition of higher priority than N_e with only finitely many requirements.
If $f \in B$ where B is a requirement for some I_{e_j}, there must be an n such that f is in the requirement for $e_j, n+1$. Again, by definition of $f(n)$ this is impossible.

This shows the claim.

We end the proof of the theorem by showing that $\{e\}(f)$ is unde-fined. If $\{e\}(f)$ were defined there would be arbitrary large s and some n such that $\{e\}_s(\bar{f}(n))\downarrow$, $s > \bar{f}(n)$.

By the construction of f , $B_{\bar{f}(n)}$ is not included in the active requirements for conditions of higher priority than N_e at any stage, so we will at some stage $s' > s$ act according to the instruction for the case $R_i = N_e$. But, either we will get a requirement for N_e active throughout or we will stop producing requirements. s was arbi-trary large, so both cases are impossible. This gives the contradiction, and the theorem is proved.

□

One of several alternative but equivalent ways of defining the recursive functions, is by adding the μ-operator to the operators gen-erating the primitive recursive functions. A very natural question is then if this will be true for Kleene's computations as well: can we re-place S9 by the μ-operator and get the same class of computable func-tions relative to a given object ?

A standard way of introducing the μ-operator is by demanding:

If $G(x,\vec{y})$ is total and recursive and

$\forall\vec{y}\ \exists x\ G(x,\vec{y}) = 0$ then

$F(\vec{y}) = \mu x(G(x,\vec{y})=0) = (\text{least } x)(G(x,\vec{y})=0)$

is recursive.

Thinking in terms of functionals, we may regard the μ-operator as a partial computable functional of type 2 by the following definition:

$\mu(f) = \text{least } x$ such that $f(x) = 0$ if such x exists

$\mu(f)$ is undefined otherwise

We may then use the following definition of μ-recursion:

Definition 4.22
Let ψ,φ be functionals. _φ is μ-recursive in ψ_ if φ is primitive recursive in ψ and μ .

Remark 4.23
If f and g are functions, then f is μ-recursive in g if and only if f is recursive in g . This is one of the basic facts about ordinary recursion theory.

Presenting μ-recursion the way we do in 4.22 we see that all computations for μ-recursion essentially have finite length (see Lemma 4.25 below). This may be used to show that μ-recursion is strictly weaker than Kleene-computability, e.g. the functions μ-recursive in ^2E are exactly the arithmetic functions (those definable by a 1st order formula over $\mathbb{N}$) while the 1-section of ^2E is exactly Δ_1^1 (Kleene [22]). But ^2E is not continuous, so we might ask: For continuous F, will μ-recursion and computability coincide ?

This was answered in the negative by J. Bergstra [1]. We will give his argument below. It was actually for this purpose he invented the functionals $F_e^{e'}$ defined in 4.14 $\underline{c}$.

Definition 4.24

Let $\vec{\phi}$ be a sequence of total functionals, $\{e\}(\vec{\phi},\mu)$ a computation. We let $h(e,\vec{\phi})$ be the maximal number of consecutive applications of S8 in $\{e\}(\vec{\phi},\mu)$ (not including applications of μ) defined by

$h(e,\vec{\phi}) = 0$ if $\{e\}(\vec{\phi},\mu)$ is an initial computation

$h(e,\vec{\phi}) = \max\{h(e',\vec{\phi}')\}$; $\{e'\}(\vec{\phi}',\mu)$ is an immediate subcomputation
$\qquad\qquad\qquad\qquad\qquad$ of $\{e\}(\vec{\phi},\mu)\}$ if e is not an index for S8
$\qquad\qquad\qquad\qquad\qquad$ applied on an element in $\vec{\phi}$

$h(e,\vec{\phi}) = \sup\{1+h(e_1,\varphi,\vec{\phi})\}$ if $\{e\}(\vec{\phi},\mu) = \varphi_1(\lambda\varphi\{e_1\}(\varphi,\vec{\phi},\mu))$

Kleene [22] called h the φ-height of a computation.

Lemma 4.25

If $\{e\}(\vec{\phi},\mu)\!\downarrow$ and the computation is generated by S1 - S8 alone, then there is a number k such that for all $\vec{\phi}'$ of the same types as $\vec{\phi}$

$\qquad\qquad \{e\}(\vec{\phi}',\mu)\!\downarrow$ iff $h(e,\vec{\phi}') \leq k$

Proof: The proof is by a trivial induction on the length of computations and is omitted.
$\qquad\qquad\qquad\qquad$ □

We are now ready to prove that μ-recursion is strictly weaker than computability also for continuous functionals. The result is due to J. Bergstra [1].

Theorem 4.26

There is a continuous functional F of type 2 and an $f \in 1\text{-sc}(F)$ such that f is not μ-recursive in F .

Proof: Let $\{e_i\}_{i \in \omega}$ be a recursive sequence of indices such that W_{e_0} is recursive and for all i

$$W_{e_i} <_T W_{e_{i+1}}$$

This is possible to construct by a standard priority argument (see e.g. Sacks [39] or Shoenfield [43]).

Let $F(i,x,f) = F_{e_i}^{e_{i+1}}(x,f)$ where $F_e^{e'}$ is as in Definition 4.14 $\underline{c}$.

The theorem is a direct consequence of the following two claims.

Claim 1

$$\text{Let} \quad f(i,x) = \begin{cases} \mu n \, T(e_i,x,n) & \text{if} \quad x \in W_{e_i} \\ 0 & \text{if} \quad x \notin W_{e_i} \end{cases}$$

Then f is computable in F.

Claim 2

Whenever g is μ-recursive in F there is an m such that g is recursive in W_{e_m}.

Proof of claim 1: Since W_{e_0} is recursive, $\lambda x \, f(0,x)$ will be recursive. Now by definition $\lambda y \, f(i,y)$ is a modulus for W_{e_i}, so

$$F_{e_i}^{e_{i+1}}(x, \lambda y \, f(i,y)) = \begin{cases} 0 & \text{if} \quad \exists n \, T(e_i,x,n) \\ 1 & \text{otherwise} \end{cases}$$

from which we easily compute $f(i+1,x)$.

So $\lambda x \, f(i+1,x)$ is uniformly computable from $F_{c_i}^{e_{i+1}}$ and $\lambda y \, f(i,y)$.

By the recursion theorem, f is clearly computable in F.

$\square$ Claim 1

Proof of claim 2: Recall the definition of $G_e^{e'}$ from the proof of Lemma 4.16 $\underline{e}$.

Let

$$H_m(i,x,f) = \begin{cases} F(i,x,f) & \text{if} \quad i < m \\ G_{e_i}^{e_{i+1}}(x,f) & \text{if} \quad m \leq i \end{cases}$$

H_m is clearly computable in W_{e_m}, and H_m is a partial subfunction of F.

Let h be as in Lemma 4.25.

<u>Subclaim</u>

If $\{e\}(\vec{n},F,\mu) \simeq k$ and $h(e,\vec{n},F) \leq m$ then

$$\{e\}(\vec{n},H_m,\mu) \simeq k$$

<u>Proof of subclaim:</u> The proof is by induction on m.
If $m = 0$, we did not consult F at all, so

$$\{e\}(\vec{n},S,\mu) \simeq k$$

for any functional S, partial or total, so assume $m > 0$.

The only non-trivial case to consider will be

$$\{e\}(\vec{n},F,\mu) \simeq F(i,x,\lambda s\{e_1\}(s,\vec{n},F,\mu))$$

where i,x occur in $\vec{n}$.

For each s, $h(e_1,s,\vec{n},F) \leq m-1$ and by the induction hypothesis

$$\lambda s\{e_1\}(s,\vec{n},F,\mu) = \lambda s\{e_1\}(s,\vec{n},H_{m-1},\mu)$$

But then

$$g = \lambda s\{e_1\}(s,\vec{n},H_{m-1},\mu)$$

is recursive in $W_{e_{m-1}}$.
If $i < m$ then $F(i,x,g) = H_m(i,x,g)$, and if $i \geq m$ then W_{e_i} is not
recursive in g, so

$$F(i,x,g) = F_{e_i}^{e_{i+1}}(x,g) = G_{e_i}^{e_{i+1}}(x,g) = H_m(i,x,g)$$

Since $H_{m-1} \subseteq H_m$ we will have that

$$g(s) = \{e_1\}(s,n,H_m,\mu)$$

for all s, and we see that

$$\{e\}(\vec{n},F,\mu) = \{e\}(\vec{n},H_m,\mu)$$

□ Subclaim

Now, if $f = \lambda x\{e\}(x,F,\mu)$ where e is an index for a primitive recursive function, there is by Lemma 4.25 a number m such that for all x

$$h(e,x,F) \leq m$$

By the subclaim we will then have that

$$f = \lambda x\{e\}(x,H_m,\mu)$$

Thus f is computable in H_m and H_m is computable in W_{e_m} so f is
recursive in W_{e_m}.
□ Claim 2

Now, let f be the function constructed in Claim 1. If f was recursive in W_{e_m} we would obtain a contradiction since all W_{e_i}, and in particular $W_{e_{m+1}}$, are recursive in f . This would lead to $W_{e_{m+1}}$ recursive in W_{e_m} which contradicts the choice of $\{e_i\}_{i \in \omega}$.

But by Claim 2 then, f cannot be μ-recursive in F .

Let f be as in Claim 1 of the last proof. Then there is an index e such that $f(x) = \{e\}(x,F)$ for all x .

By inspection of the algorithm for f we see that e may be chosen such that for all y,G

$$\{e\}(y,G)\!\downarrow$$

Clearly $\Phi(y,G) = \{e\}(y,G)$ cannot be μ-recursive. Thus we may draw the following conclusion.

Corollary 4.27

There is a computable countable functional of type 3 that is not μ-recursive.

Remark 4.28

<u>a</u>   If we look at the argument showing that μ-recursion is strictly weaker than computability when relativized to ²E , we may construct a computable type 3 functional in the maximal type structure that is not μ-recursive. It is not obvious and probably not true that this functional will still not be μ-recursive if we restrict it to the continuous objects.

<u>b</u>   If we look at the proof of claim 1 , we see that f(i,x) may be defined from F by induction on i . This definition does not fit into S5 , since we use S8 at each stage of the induction, and as we have seen, such inductions may lead outside the class of primitive recursive functions.

4.3   The fan-functional

In section 4.2 we showed that there are irreducible functionals of type 2. Since any continuous functional of type 2 is computable in any-one of its associates, this must mean that we cannot always find associates for F computable in F .

If we try to compute a type 3 functional φ from an associate α for φ , we see that in order to compute φ(F) the most natural attempt would be to look for an associate β for F , but by the results of section 4.2 there is no hope of success.

Based on such considerations, we should have a feeling that there may be non-obtainable functionals of type 3 , and indeed we will construct one.

We will give an example due to Tait. We will give two proofs, both in the line of Gandy-Hyland [13], because we think that both proofs are informative.

Tait's functional is defined through an analysis of the behaviour of the continuous functionals of type 2 on compact subsets of Tp(1) . In section 7.1 we will construct non-obtainable functionals of type $\geqslant$ 3.

Definition 4.29

<u>a</u>   Let $A \subseteq Tp(1)$. A is said to be <u>bounded</u> if there is an $f \in Tp(1)$

    such that $\forall g \in A \; \forall n \; g(n) \leq f(n)$

<u>b</u>   Let $f \in Tp(1)$. Let $C_f = \{g \in Tp(1); \forall n \, g(n) \leq f(n)\}$.

Remark 4.30

By 'bounded' we here mean 'bounded in all coordinates n '.

The following lemma will be well-known to most readers, but we include it for the sake of completeness.

Lemma 4.31

Let $K \subseteq Tp(1)$. Then K is compact if and only if K is both closed and bounded.

<u>Proof</u>: Let K be compact. Then K is closed. For each n , let $0_{k,n} = \{f \; ; \; f(n) = k\}$.

$0_{k,n}$ is open and $\{0_{k,n}\}_{k \in \omega}$ is a covering of Tp(1) .

By compactness there is a k_n such that $K \subseteq 0_{1,n} \cup \ldots \cup 0_{k_n,n}$.

Let $f(n) = k_n$. Then K is bounded by f .

To show the other direction, it is sufficient to prove that each C_f is compact. But C_f is essentially the product of the finite discrete sets $\{\{0,\ldots,f(n)\}\}_{n \in \omega}$ with the product topology, and C_f will then be compact.   ☐

Remark 4.32

The reader is wellcome to give a more 'constructive' proof.

Lemma 4.33

Let $f:\omega \to \omega$ and let $F:\mathrm{Tp}(1) \to \omega$ be continuous. Then there is a number n such that

$$\forall g,h \in C_f \; (\bar{g}(n)=\bar{h}(n) \to F(g)=F(h))$$

Proof: Let $B = \{B_\tau ; F \text{ is constant on } B_\tau\}$. B covers $\mathrm{Tp}(1)$. Since C_f is compact there are $B_{\tau_1},\ldots,B_{\tau_k} \in B$ such that $C_f \subseteq B_{\tau_1} \cup \ldots \cup B_{\tau_k}$. Let $n = \max(\mathrm{lh}(\tau_1),\ldots,\mathrm{lh}(\tau_k))$. Then n clearly works.

$\square$

Remark 4.34

The compactness of C_f clearly has an equivalent formulation in the Brouwer fan theorem:

Any wellfounded tree with finite branching
is finite.

Lemma 4.33 is just a special case of the same result. It is through this connection that Tait's functional carries its name.

An n with the property of Lemma 4.33 is called a modulus of uniform continuity for F over C_f.

Definition 4.35

By the fan-functional Φ we mean

$$\Phi(F,f) = \mu n \; \forall g,h \in C_f (\bar{g}(n)=\bar{h}(n) \to F(g)=F(h))$$

If we just look at the definition, the fan-functional seems rather complicated, $\Phi(F,f)$ depends on F applied to all members of the often uncountable set C_f. But as our next result shows, it is not complicated at all.

Theorem 4.36

The fan-functional Φ is recursive.

Proof: First we show how to compute $\Phi(F,f)$ from f and an associate α for F. Then we construct a recursive associate for Φ.
So let f and $\alpha \in \mathrm{As}(F)$ be given. By the same argument as in Lemma 4.32 we see that there is an n such that

$$\forall g \in C_f \; \alpha(\bar{g}(n)) > 0$$

It will be recursive in f,α to decide if n has this property, since it is equivalent to

$$\forall \tau \, (lh(\tau) = n \wedge \forall i < n \, \tau(i) \leq f(i) \rightarrow \alpha(\tau) > 0)$$

which clearly is recursive in f, α since there are only finitely many τ to consider.

Pick n_0 minimal such that

$$\forall g \in C_f \, \alpha(\bar{g}(n_0)) > 0$$

Let

$$C_\sigma = \{\tau \, ; \, lh(\tau)=lh(\sigma) \wedge \forall i<lh(\sigma)(\tau(i)\leq\sigma(i))\}$$

Let $\varphi : C_{\bar{f}(n_0)} \rightarrow \omega$ be defined by

$$\varphi(\tau) = \alpha(\tau)-1$$

Clearly, for any $g \in C_f$ we have $\varphi(\bar{g}(n_0)) = F(g)$. $C_{\bar{f}(n_0)}$ and φ are finite and uniformly recursive in α, f and n_0 .

Let $n \leq n_0$ be minimal such that

$$\forall \tau_1 \tau_2 \in C_{\bar{f}(n_0)} \, (\bar{\tau}_1(n)=\bar{\tau}_2(n) \rightarrow \varphi(\tau_1)=\varphi(\tau_2))$$

Then $n = \Phi(F,f)$ and we have found n recursively in f, α .

In order to construct an associate for Φ we will just see which parts of α and f we needed in the algorithm above. We define an associate for Φ below, and leave the proof for the reader.

$$\beta(\sigma_1,\sigma_2) = \begin{cases} n+1 & \text{if } \forall \tau \in C_{\sigma_2}(\sigma_1(\tau)>0) \wedge n \leq lh(\sigma_2) \text{ is minimal} \\ & \text{such that } \forall \tau_1 \tau_2 \in C_{\sigma_2}(\bar{\tau}_1(n)=\bar{\tau}_2(n) \rightarrow \sigma_1(\tau_1)=\sigma_2(\tau_2)) \\ \\ 0 & \text{if there is no such } n \end{cases}$$

□

From now on and throughout this section we will let Φ be the fan-functional. We will show that Φ is not computable. The proof may easily be relativized to any $g \in Tp(1)$, and we will conclude that Φ is not 2-obtainable, and hence Φ is non-obtainable.

The following lemma is a result in ordinary recursion theory.

Lemma 4.37

Let $A_0 = \{e;\{e\}(e) = 0\}$
$A_1 = \{e;\{e\}(e) = 1\}$

A_0 and A_1 cannot be separated by any recursive set B .

Proof: Assume that $A_1 \subseteq B \subseteq A_2$ where B is recursive, and let

$$\lambda x\{e\}(x)$$

be the characteristic function of B. We then have

$$e \in B \leftrightarrow \{e\}(e) = 1 \rightarrow e \in A_2 \rightarrow e \notin B$$
$$e \notin B \leftrightarrow \{e\}(e) = 0 \rightarrow e \in A_1 \rightarrow e \in B$$

so $e \in B \leftrightarrow e \notin B$ which is impossible.

□

Now define

$$R(\tau) \leftrightarrow \exists i,j < lh(\tau)(\{i\}(i)\downarrow \text{ in at most } j \text{ steps and}$$
$$\tau(i) = 1 \dot{-} \{i\}(i) \ (\text{mod } 2))$$

Lemma 4.38

<u>a</u>   If f is recursive, there is an n such that $R(\bar{f}(n))$

<u>b</u>   There is an $f : \omega \rightarrow \{0,1\}$ such that $\forall n \neg R(\bar{f}(n))$.

Proof:

<u>a</u>   Assume $\forall n \neg R(\bar{f}(n))$. Let $B = \{n; f(n) \equiv 0 \ (\text{mod } 2)\}$

If f is recursive then B will be recursive, so by Lemma 4.37
it is sufficient to show that B separates A_0 and A_1 . Let $i \in A_0$.
Choose $n > i$ such that $\{i\}_n(i) = 0$.
Since $\bar{f}(n)$ is not in R we must have

$$f(i)(\text{mod } 2) \neq 1 \dot{-} \{i\}(i) = 1$$

so $f(i) \equiv 0 \ (\text{mod } 2)$ and $i \in B$.
If $i \in A_1$, choose $n > i$ such that $\{i\}_n(i) = 1$. Again, since $\bar{f}(n)$
is not in R we must have

$$f(i)(\text{mod } 2) \neq 1 - \{i\}(i) = 0$$

so $i \notin B$.
This shows that B separates A_0 and A_1 .

<u>b</u>   Let

$$f(e) = \begin{cases} 0 & \text{if } \{e\}(e) \equiv 0 \ (\text{mod } 2) \\ 1 & \text{otherwise} \end{cases}$$

f clearly satisfies part b.

□

Now, choose e such that W_e is not recursive.
Let

$$M_e(x,f) = \begin{cases} 0 & \text{if } \exists n (T(e,x,n) \wedge \neg R(\bar{f}(n))) \\ 1 & \text{otherwise} \end{cases}$$

Lemma 4.39

<u>a</u> M_e is computable in W_e .

<u>b</u>   There is a partial computable subfunction $G_e \subseteq M_e$ such that $G_e(x,f)$ is defined for all recursive f .

<u>c</u> W_e is computable in M_e and Φ .

Proof: We leave <u>a</u> and <u>b</u> for the reader and concentrate on <u>c</u>. Given x, let $M_{e,x}(f) = M_e(x,f)$. By Lemma 4.38 <u>b</u> and the definition of M_e we see that

$$x \in W_e \leftrightarrow \exists f \in C_h M_{e,x}(f) = 0$$

where h is the constant 1 function.
Let $n = \Phi(M_{e,x},h)$.
Let $g_1,\ldots,g_{2n}$ be recursive functions such that $\forall f \in C_h \exists i \leq 2^n (\bar{f}(n) = \bar{g}_i(n))$ (in which case $M_{e,x}(f) = M_{e,x}(g_i)$). So

$$x \in W_e \leftrightarrow \exists i \leq 2^n M_{e,x}(g_i) = 0$$

but this is computable from M_e and n , and n was computed from x, M_e and Φ , so W_e is computable in M_e, Φ . $\square$

The following theorem was first proved by Tait who never published it. See Gandy-Hyland [13] for further references.

Theorem 4.40
   The fan-functional Φ is not computable.

Proof: If Φ is computable, we will have that

$$1\text{-sc}(M_e) = 1\text{-sc}(M_e,\Phi)$$

But by Lemma 4.39 <u>b</u> and theorem 4.11, $1\text{-sc}(M_e)$ consists just of the recursive functions while $W_e \in 1\text{-sc}(M_e,\Phi)$, so this is impossible. $\square$

We have now established that Φ is not computable in any type 2 functional. We will now show that for any continuous F of type 2 there is a functional H_F of type 2 uniformly computable in Φ, F , such that

$$1\text{-sc}(\Phi,F) = 1\text{-sc}(H_F)$$

This was proved in Bergstra [1], and we will more or less follow his argument.

To carry through this argument, it is better to regard a functional ϕ^* equivalent to ϕ .

Definition 4.41

<u>a</u>   Let σ be a sequence-number, $f:\omega \to \omega$.
Let
$$C_{\sigma,f} = C_f \cap B_\sigma .$$

<u>b</u>   If $F \in Ct(2)$, let
$$D_{F,f,\sigma} = \Sigma\{2^{n+1}; \text{ for some } g \in C_{\sigma,f} (F(g)=n)\}$$

If $C_{\sigma,f} = \emptyset$ we let $D_{F,f,\sigma} = 0$.

<u>c</u> $\phi^*(F,f,\sigma) = D_{F,f,\sigma}$.

Remark 4.42

$D_{F,f,\sigma}$ is coding the finite set of values of F on $C_{\sigma,f}$.

Lemma 4.43

ϕ and ϕ^* are equivalent.

Proof:

<u>i</u> ϕ is computable in ϕ^*: To compute $\phi(F,f)$ find the least n such that for all $\sigma \in C_{\bar{f}(n)}$ we have that $\phi^*(F,f,\sigma)$ codes a singleton, i.e. $\phi^*(F,f,\sigma)$ is of the form 2^k for some $k \geq 0$.

<u>ii</u> ϕ^* is computable in ϕ: To compute $\phi^*(F,f,\sigma)$, check if $\forall i < lh(\sigma)(\sigma(i) \leq f(i))$. If this is not the case we have that $C_{\sigma,f} = \emptyset$ and $\phi^*(F,f,\sigma) = 0$. If this is the case, let $n = \phi(F,f)$. Then
$$D_{F,f,\sigma} = \Sigma\{2^m; \exists \tau \in C_{\bar{f}(n)} (\tau \text{ is compatible with } \sigma \text{ and } F(\tau^\frown 0)=m)\}$$
where $\tau^\frown 0$ is the function which begins like τ and is zero for larger arguments.

□

Definition 4.44

Let $H_F(f,\sigma) = \phi^*(F,f,\sigma)$

Clearly H_F is uniformly computable in ϕ^* , F so we obtain

Lemma 4.45

$1\text{-sc}(H_F) \subseteq 1\text{-sc}(\phi^*,F)$.

□

We are actually going to show that the two 1-sections are the same. To do this we want to simulate all computations in ϕ, F as computations in H_F. First we will show that ϕ is actually computable when applied to a functional computable in an H_F.

Lemma 4.46

There is a partial computable functional ψ_0 such that whenever σ is a sequence, $f:\omega \to \omega$, $\vec{m}$ are number parameters, $C_{\sigma,f} \neq 0$ and

$$\lambda\beta\{e\}(\beta,\vec{m},\vec{\alpha},H_F) \text{ is total on } C_{\sigma,f}$$

then $\psi_0(e,\vec{m},\vec{\alpha},\sigma,f,H_F)$ is a modulus of uniform continuity for $\lambda\beta\{e\}(\beta,\vec{m},\vec{\alpha},H_F)$ on $C_{\sigma,f}$.

Proof: We will construct ψ_0 using the recursion theorem. The argument will be by induction on the following structure $<U_F,<_F>$ where

$$U_F = \{<e,\vec{m},\sigma>\; ; \lambda\beta\{e\}(\beta,\vec{m},\vec{\alpha},H_F) \text{ is total on } C_{\sigma,f} \neq \emptyset\} \; .$$

$<e,\vec{m},\sigma> <_F <d,\vec{n},\tau>$ if τ is a subsequence of σ and

$$\forall\beta \in C_{\sigma,f}(\{e\}(\beta,\vec{m},\vec{\alpha},H_F) \text{ is a subcomputation of } \{d\}(\beta,\vec{n},\vec{\alpha},H_F)) \; .$$

Claim

$<_F$ is a transitive well-founded relation.

Proof: Transitivity follows from the transitivity of the subcomputation-relation, and the fact that

$$<e,\vec{m},\sigma> <_F <d,\vec{n},\tau> \to C_{\sigma,f} \subseteq C_{\tau,f} \; .$$

Assume that $<_F$ is not well-founded and let $\{<e_i,\vec{m}_i,\sigma_i>\}_{i\in\omega}$ be a descending sequence. Choose $\beta \in \bigcap_{i\in\omega} C_{\sigma_i,f}$. Then, by the definition of U_F and $<_F$ $\{\{e_i\}(\beta,\vec{m}_i,\vec{\alpha},H_F)\}_{i\in\omega}$ is a descending sequence of computations, contradicting that the subcomputation-relation is well-founded.

$\quad\quad\quad\quad\quad\quad\quad\quad\quad\quad\quad\quad\quad\quad\quad\quad\quad\quad\quad$ □ Claim

We will now construct ψ_0, and at the same time we will show by induction on $<_F$ that if $<e,\vec{m},\sigma> \in U_F$, then $\psi_0(e,\vec{m},\vec{\alpha},\sigma,f)$ is defined and has the value described in the lemma. We will split the construction into nine cases following S1 - S9.

Cases 1, 2 and 3 are trivial, we let $\psi_0(e,\vec{m},\vec{\alpha},\sigma,f,H_F) = 0$.

Case 4 $\{e\}(\beta,\vec{m},\vec{\alpha},H_F) = \{e_1\}(\beta,\{e_2\}(\beta,\vec{m},\vec{\alpha},H_F),\vec{m},\vec{\alpha},H_F)$.

Let $t_1 = \max\{\psi_0(e_2,\vec{m},\vec{\alpha},\sigma,f,H_F),\mathrm{lh}(\sigma)\}$. Let $\tau_1,\ldots,\tau_k$ be all extensions of σ of length t_1 that are bounded by f (if $t_1 = \mathrm{lh}(\sigma)$ there will be at most one $\tau_1 = \sigma$). Let $n_1,\ldots,n_k$ be the respective constant values of $\lambda\beta\{e_2\}(\beta,\vec{m},\vec{\alpha},H_F)$ on $C_{\tau_1,f},\ldots,C_{\tau_k,f}$.

Let $t_2 = \max\{t_1,\psi_0(e_1,n_i,\vec{m},\vec{\alpha},\tau_i,f,H_F); i \le k\}$. Note that $<e_1,n_i,\vec{m},\tau_i> <_F <e,\vec{m},\sigma>$ so by the induction hypothesis ψ_0 should work here. But then we let

$$\psi_0(e,\vec{m},\vec{\alpha},\sigma,f,H_F) = t_2$$

We must show that this works: Let $\gamma \in C_{\sigma,f}$, $\beta \in C_{\sigma,f}$ and assume that $\bar{\gamma}(t_2) = \bar{\beta}(t_2)$. Since $t_2 \ge t_1$ there will be an $i \le k$ such that $\alpha,\beta \in C_{\tau_i,f}$ and

$$\{e_2\}(\gamma,\vec{m},\vec{\alpha},H_F) = \{e_2\}(\beta,\vec{m},\vec{\alpha},H_F) = n_i$$

But since $t_2 \ge \psi_0(e_1,n_i,\vec{m},\vec{\alpha},\tau_i,f,H_F)$ we will have that

$$\{e_1\}(\gamma,n_i,\vec{m},\vec{\alpha},H_F) = \{e_1\}(\beta,n_i,\vec{m},\vec{\alpha},H_F)$$

showing that

$$\{e\}(\gamma,\vec{m},\vec{\alpha},H_F) = \{e\}(\beta,\vec{m},\vec{\alpha},H_F)$$

This ends case 4.

Case 7 (Special case)

$$\{e\}(\beta,\vec{m},\vec{\alpha},H_F) = \beta(m)$$

Let $\psi_0(e,\vec{m},\vec{\alpha},\sigma,f,H_F) = m+1$.

Case 8 $\{e\}(\beta,\vec{m},\vec{\alpha},H_F) = H_F(\lambda x\{e_1\}(\beta,x,\vec{m},\vec{\alpha},H_F),\tau)$. Let $\delta(x) = \max\{\{e_1\}(\beta,x,\vec{m},\vec{\alpha},H_F); \beta \in C_{\sigma,f}\}$.

By the induction hypothesis $\psi_0(e_1,x,\vec{m},\sigma,f,H_F)$ are defined and work well for all x . It follows that δ will be computable in the parameters.

If for some i we have that $\tau(i) > \delta(i)$ we know that $H_F(\lambda x\{e_1\}(\beta,x,\vec{m},\vec{\alpha},H_F),\tau) = 0$ for all $\beta \in C_{\sigma,f}$, and we may set $\psi_0(e,\vec{m},\vec{\alpha},\sigma,f,H_F) = 0$. So assume that for all $i < \mathrm{lh}(\tau)$ we have $\tau(i) \le \delta(i)$.

We know that F has a modulus of uniform continuity over $C_{\tau,\delta}$, and for any extension π of τ , F is constant on $C_{\pi,\delta}$ iff $H_F(\delta,\pi)$ is a singleton. So choose $m \ge \mathrm{lh}(\tau)$ such that for all $\pi \in C_{\bar{\delta}(m)}$, if π extends τ then $H_F(\delta,\pi)$ is a singleton (i.e. on the form 2^k) and let

$$\psi_0(e,\vec{m},\vec{\alpha},\sigma,f,H_F) = \max\{\psi_0(e_1,x,\vec{m},\vec{\alpha},\sigma,f,H_F); \; x \leq m\} = n$$

We end the proof of case 8 by showing that our choice of n will work.

So assume $\vec{\beta}(n) = \vec{\gamma}(n)$, β and γ both in $C_{\sigma,f}$. Then for each $x \leq m$

$$\{e_1\}(\beta,x,\vec{m},\vec{\alpha}) = \{e_1\}(\gamma,x,\vec{m},\vec{\alpha})$$

Let $\pi = \lambda x < m\{e_1\}(\beta,x,\vec{m},\vec{\alpha})$

$\quad \pi_\beta = \lambda x\{e_1\}(\beta,x,\vec{m},\vec{\alpha})$

$\quad \pi_\gamma = \lambda x\{e_1\}(\gamma,x,\vec{m},\vec{\alpha})$

We must show that

$$H_F(\pi_\beta,\tau) = H_F(\pi_\gamma,\tau)$$

If for some $i < lh(\tau)$ $\pi(i) < \tau(i)$, both values are 0. So assume that $\forall i < lh(\tau)$ $\tau(i) \leq \pi(i)$. The common value of $H_F(\pi_\beta,\tau)$ and $H_F(\pi_\gamma,\tau)$ will then be

$$\Sigma\{2^s; \; \exists\pi' \in C_\pi \;(\pi' \text{ extends } \tau \text{ and } H_F(\delta,\pi') = 2^s)\}$$

This ends the proof of case 8.

The rest of the cases are easy and omitted.

$\square$

Lemma 4.47

There is a partial recursive function ρ such that whenever $\{e\}(\vec{m},\vec{\alpha},F,\Phi)\simeq k$ then $\{\rho(e)\}(\vec{m},\vec{\alpha},H_F)\simeq k$.

Proof: This is a standard reindexing argument, we define $\rho(e)$ by cases 1 - 9. The only difficult cases are applications of F or Φ.

To cover the application of F we just show that F is computable in H_F. To compute $F(f)$ look for the least n such that $H_F(f,\bar{f}(n))$ is on the form 2^s. Then $F(f) = s$. So there is one case left to consider:

$$\{e\}(\vec{m},\vec{\alpha},F,\Phi) = \Phi(\lambda\beta\{e_1\}(\vec{m},\beta,\vec{\alpha},F,\Phi),\lambda x\{e_2\}(x,\vec{m},\vec{\alpha},F,\Phi))$$

$$= \Phi(\lambda\beta\{\rho(e_1)\}(\vec{m},\beta,\vec{\alpha},H_F),\lambda x\{\rho(e_2)\}(x,\vec{m},\vec{\alpha},H_F))$$

Let $\delta = \lambda x\{\rho(e_2)\}(x,\vec{m},\vec{\alpha},H_F)$. Let ψ_0 be as in Lemma 4.46 and let $m = \psi_0(\rho(e_1),\vec{m},\vec{\alpha},<>,\delta,H_F)$. m will be one modulus of uniform continuity, and we have seen before how to find the least one, m_0 from m. Let $\{\rho(e)\}(\vec{m},\vec{\alpha},H_F) = m_0$.

$\square$

We may now conclude

Theorem 4.48

Let $F \in Ct(2)$, Φ as before. Let H_F be as in Definition 4.44. Then

$$1\text{-}sc(H_F) = 1\text{-}sc(F,\Phi)$$

Proof: By Lemma 4.45 we have $1\text{-}sc(H_F) \subseteq 1\text{-}sc(F,\Phi)$ and by the reindexing argument of Lemma 4.47 we will have $1\text{-}sc(F,\Phi) \subseteq 1\text{-}sc(H_F)$.

□

Corollary 4.49

Let $F_e^{e'}$ be as in Definition 4.14 $\underline{c}$ and assume that W_e is not recursive. Then all elements of $1\text{-}sc(F_e^{e'},\Phi)$ are recursive.

Proof: By Theorem 4.48 it is sufficient to show that $1\text{-}sc(H_{F_e^{e'}})$ only contains recursive elements.

By Theorem 4.11 it is sufficient to find a partial computable subfunctional of $H_{F_e^{e'}}$ defined for all recursive arguments.

Note that if f is not a modulus for W_e then there are no moduli for W_e in C_f. So let f be recursive. $H_{F_e^{e'}}(x,f,\sigma)$ is the coded set of values of $F_e^{e'}(x,\beta)$ for $\beta \in C_{\sigma,f}$. But to find this set, pick first the least n such that $\neg Mod(\bar{f}(n))$. If there is no $m < n$ such that $T(e,x,m)$ we only have 1 in this set, so $H_{F_e^{e'}}(x,f,\sigma) = 2^1 = 2$.

If there is such m, we get 0 in the set, and if for some $\pi \in C_{\bar{f}(n)}$ we have $\neg Mod(\bar{\pi}(m))$ we also get 1 in the set, so the value of $H_{F_e^{e'}}$ is either 1 or 3, and we may effectively decide which to choose. Above we have described an algorithm for computing $H_{F_e^{e'}}(x,f,\sigma)$ that will work for all recursive f. We may then use Theorem 4.11.

□

Remark 4.50

This corollary shows that $F_e^{e'}$ is irreducible modulo Φ. In the beginning of this section we thought that since there are irreducible functionals of type 2 there ought to be non-obtainable functionals of type 3. By Corollary 4.49 we might think that there are functionals of type 3 non-obtainable relative to Φ. In the next section we will show that this is the case. We will also by an alternative proof show once more that Φ is not obtainable.

4.4 The Γ-functional

The fan-functional Φ was based on compactness. In this section we will investigate another functional, Γ , defined by Gandy and studied in Gandy-Hyland [13]. Gandy proved that Γ has a recursive associate, while Hyland proved that Γ is not computable in Φ and any function $f: \omega \to \omega$.

Definition 4.51

<u>a</u>   Let $F \in Ct(2)$, $f \in Tp(1)$, τ a sequence

Let $F_\tau(f) = F(\tau^\wedge f)$

where

$$(\tau^\wedge f)(i) = \begin{cases} \tau(i) & \text{if } i < \mathrm{lh}(\tau) \\ f(i - \mathrm{lh}(\tau)) & \text{if } i \geq \mathrm{lh}(\tau) \end{cases}$$

<u>b</u>   Let

$$\Gamma(F) = \begin{cases} k & \text{if } F \text{ is constant } k \\ F_{<0>}(\lambda n \Gamma(F_{<n+1>})) & \text{if } F \text{ is not constant} \end{cases}$$

Lemma 4.52

Γ is defined for all F and has a recursive associate.

<u>Proof</u>: That $\Gamma(F)$ is defined is straightforward by induction on the Kalmar-rank of F (Definition 4.8).

Now, let α be an associate for F . The rank of α will be the ordinal height of the well-founded tree

$$T_\alpha = \{\sigma ; \alpha(\sigma) = 0\} .$$

By induction on the rank of α we show that uniformly recursive in α we can find a finite bit $\bar{\alpha}(t)$ deciding $\Gamma(F)$.

If α has rank 0 , i.e. if $\alpha(<>) = k+1$ for some k , then F is constant k and $\Gamma(F) = k$.

So, if $\alpha(<>) = k+1$ let $\beta(\bar{\alpha}(t)) = k+1$ for all t such that $<> < t$ ($< >$ is the sequence number of the empty sequence).

If $\alpha(<>) = 0$, we let $\alpha_n(\tau) = \alpha(n^\wedge \tau)$ for all τ . α_n will be an associate for $F_{<n>}$, and by the induction hypothesis there are t_n recursive in α_n, n such that $\bar{\alpha}_n(t_n)$ decides $\Gamma(F_{<n>})$. But then $\delta(n) = \Gamma(F_{<n+1>})$ is recursive in α .

Choose s such that $\alpha_0(\bar{\delta}(s)) > 0$. Then, to decide $\Gamma(F)$ it is sufficient to know $\Gamma(F_{<1>}), \ldots, \Gamma(F_{<s>})$.

These values are given by $\bar{\alpha}_1(t_1), \ldots, \bar{\alpha}_s(t_s)$, so we choose t such that $\bar{\alpha}(t)$ contains enough information about α to compute $\alpha_0(\bar{\delta}(s)), \bar{\alpha}_1(t_1), \ldots, \bar{\alpha}_s(t_s)$, and we let

$$\beta(\bar{\alpha}(n)) = \begin{cases} k+1 & \text{if } n \geq t \text{ and } \alpha_0(\bar{\delta}(s)) = k+1 \\ 0 & \text{otherwise} \end{cases}$$

□

This shows that Γ is recursive. Before we can show that Γ is not computable in Φ , we need a general definition and a general theorem.

Definition 4.53

Let $F \in \text{Ct}(2)$, $\alpha \in \text{Tp}(1)$. α is called a <u>restricted</u> <u>associate</u> for F if

<u>i</u> $\forall f \in 1\text{-sc}(F) \ \exists n \ \alpha(\bar{f}(n)) > 0$

<u>ii</u> $\forall f \ \forall n \ \forall m \geq n \ [\alpha(\bar{f}(n)) > 0 \rightarrow \alpha(\bar{f}(m)) = F(f)+1]$.

Theorem 4.54

If α is a restricted associate for F and $\{e\}(F,\vec{n}) = k$, then there will be sequences $\pi_1, \ldots, \pi_n$ such that $\alpha(\pi_1) > 0 \wedge \ldots \wedge \alpha(\pi_n) > 0$ and for all G , if G coincides with F on $\underset{i \leq n}{U} B_{\pi_i}$ and $\{e\}(G,\vec{n})\downarrow$, then $\{e\}(G,\vec{n}) = k$.

<u>Proof</u>: This is shown by a trivial induction on the computations in F , noticing that F is only applied on elements in $1\text{-sc}(F)$.

□

Remark 4.55

This result enables us to use descriptions of F that do not contain total information about F when we study computations in F . Let us see how we can use Theorem 4.54 to prove once more that Φ is not computable.

Assume that Φ is computable, and let h be the constant 1 function. Then there will be an index e such that for all F

$$\{e\}(F) = \Phi(F,h)$$

In particular $\{e\}(^2 0) = 0$. Let $f \in C_h$, f not recursive.

Let $\alpha(\tau) = \begin{cases} 1 & \text{if } \exists t \ (\tau(t) \neq f(t)) \\ 0 & \text{otherwise} \end{cases}$

Then α is a restricted associate for $^2 0$ and $\forall n \ \alpha(\bar{f}(n)) = 0$. By Theorem 4.54 there will be $\pi_1, \ldots, \pi_k$ such that

$\alpha(\pi_i) = 1$ $(i=1,\ldots,k)$ and if G coincide with ${}^2 0$ on $B_{\pi_1} U \ldots U B_{\pi_k}$ then $\{e\}(G) = 0$.

Since $f \notin B_{\pi_1} U \ldots U B_{\pi_k}$, there will be an $n > 0$ such that $B_{\bar{f}(n)} \cap (B_{\pi_1} U \ldots U B_{\pi_k}) = \emptyset$.

Define G by

$$G(g) = \begin{cases} 1 & \text{if } \bar{g}(n) = \bar{f}(n) \\ 0 & \text{if } \bar{g}(n) \neq \bar{f}(n) \end{cases}$$

Then $\phi(G,h) = n$, while $\{e\}(G) = 0$, contradiction.

We will now through some lemmas show that Γ is not 1-obtainable modulo ϕ. The argument is due to M. Hyland and is given in Gandy-Hyland [13].

Lemma 4.56

Let $\delta: \omega \to \omega$ be such that $\delta(k) > 0$ for all k.
Let $A_\delta = \{\sigma;\ \exists n(\sigma(n) < \delta(n))$ and for the least such n $(\text{lh}(\sigma) > n + \delta(n))\}$.
Let $\{\sigma_1, \ldots, \sigma_k\}$ be a finite set from A_δ, $s_1, \ldots, s_k$ fixed numbers.
For any number s there will be a functional F such that $\Gamma(F) = s$ and F is constant s_i on B_{σ_i} $(i=1,\ldots,k)$.

Proof: The proof goes by induction on k. If $k = 0$, the finite set is empty and we may choose any F such that $\Gamma(F) = s$. So assume that $k > 0$ and assume that the lemma holds for smaller sets from any $A_{\delta'}$.
Let $\delta^+(k) = \delta(k+1)$. Note that

$$\tau \in A_{\delta^+} \leftrightarrow \delta(0)^\frown \tau \in A_\delta.$$

We will use the induction-hypothesis on A_{δ^+}.

Case 1

There is no $i \leq k$ such that $\sigma_i(0) = 0$.
We may then take F constant on $B_{\sigma_1}, \ldots, B_{\sigma_k}$ such that F is constant s on $B_{<0>}$. Then $\Gamma(F) = s$.

Case 2

There is at least one $i \leq k$ such that $\sigma_i(0) = 0$.
For all such i we know that $\text{lh}(\sigma_i) > 0 + \delta(0) = \delta(0)$.
Let $s' = \text{Max}\{\sigma_i(\delta(0)); \sigma_i(0)=0\}+1$. Now $\{j; \sigma_j(0)=\delta(0)\}$ has cardinality $< k$ so by the induction hypothesis there is a functional $F_{\delta(0)}$ such that $\Gamma(F_{\delta(0)}) = s'$ and if $\sigma_j = \delta(0)^\frown \tau$ then $F_{\delta(0)}$ is constant s_j on B_τ.

Define F by

$$F(f) = \begin{cases} F_{\delta(0)}(f^+) & \text{if } f(0) = \delta(0) \\ s_j & \text{if for some } n\, \bar{f}(n) = \sigma_j \\ s & \text{if } f(0) = 0 \text{ and } f(\delta(0)) = s' \\ 0 & \text{otherwise} \end{cases}$$

It is easy to see that F is well-defined.
Let $f(0) = 0$, $f(n+1) = \Gamma(F_{<n+1>})$. Then $\Gamma(F) = F(f)$. But
$f(\delta(0)) = \Gamma(F_{<\delta(0)>}) = \Gamma(F_{\delta(0)}) = s'$ so $F(f) = s$. □

Remark 4.57

We see that there will be a lot of finite sets $\sigma_1,\ldots,\sigma_k$ such
that the value of F on $B_{\sigma_1} \cup \ldots \cup B_{\sigma_k}$ has no effect on $\Gamma(F)$. We
will use this to show that Γ is not computable in Φ .

Now, let F be given, $\{\alpha_i\}_{i\in\omega}$ be an enumeration of $1\text{-sc}(F,\Phi)$
and define δ by $\delta(i) = \alpha_i(i)+1$.
Let $\sigma \in A_{F,\tau}$ if $\forall \pi \in C_{\tau,\sigma}$ ($\pi \in A_\delta$ and F are constant on B_π) where
$C_{\tau,\sigma} = \{\pi; \tau \prec \pi \wedge \text{lh}(\pi) = \text{lh}(\sigma) \wedge \forall i < \text{lh}(\sigma)(\pi(i) \leq \sigma(i))\}$.
Recall the definition of H_F from 4.44.

Lemma 4.58

<u>a</u>   If $\sigma \in A_{F,\tau}$ then $\lambda f H_F(f,\tau)$ is constant on B_σ ,
<u>b</u>   If $f \in 1\text{-sc}(H_F)$ then there is an n such that

$$\bar{f}(n) \in A_{F,\tau}$$

Proof:

<u>a</u>   Assume that for all $\pi \in C_{\tau,\sigma}$ F is constant on B_π , and let D
be (a coded version of) the finite set of these constant values.
Let f extend σ . It is sufficient to show that $D = H_F(f,\tau)$.
If $g \in C_{\tau,f}$, there will be a $\pi \in C_{\tau,\sigma}$ such that $g \in B_\pi$, so
$H_F(f,\tau) \subseteq D$.
On the other hand, if $\pi \in C_{\tau,\sigma}$, then $\pi^\frown 0 \in B_\pi \cap C_{\tau,f}$, so the constant
value of F on B_π must be an element of $H_F(f,\tau)$. Thus
$H_F(f,\tau) = D$.

<u>b</u>   Let $f \in 1\text{-sc}(H_F)$. By the choice of δ above, just after Remark
4.57, there is a least n_0 such that $f(n_0) < \delta(n_0)$.
Let $n_1 = \max\{n+\delta(n); n \leq n_0\}$. Then for all $n > n_1$ we will have
$\bar{f}(n) \in A_\delta$, and moreover

$$\forall \pi \in C_{\bar{f}(n)} (\pi \in A_\delta)$$

By the compactness of $C_{\tau,f}$ we find $n > n_1$ such that F is constant on B_π for all $\pi \in C_{\tau,\bar{f}(n)}$. Then, by definition of $A_{F,\tau}$ we have $\bar{f}(n) \in A_{F,\tau}$. □

Lemma 4.59

Let

$$\alpha_\tau(\sigma) = \begin{cases} 0 & \text{if } \sigma \notin A_{F,\tau} \text{ or } \lambda f H_\Gamma(f,\tau) \text{ is not constant on } B_\sigma \\ k+1 & \text{if } \lambda f H_\Gamma(f,\tau) \text{ is constant } k \text{ on } B_\sigma \end{cases}$$

Then α_τ is a restricted associate for $\lambda f H_\Gamma(f,\tau)$.

Proof: Directly from Lemma 4.58. □

Lemma 4.60

If F is a continuous functional of type 2 and $\{e\}(F,\Phi) \simeq k$ then there is a $G \in Ct(2)$ such that if $\{e\}(G,\Phi)\downarrow$ then $\{e\}(G,\Phi) \simeq k$ but $\Gamma(F) \neq \Gamma(G)$.

Proof: Let α_τ be as in Lemma 4.59. By Lemma 4.47 there is a recursive function ρ such that for all $G \in Ct(2)$

$$\{e\}(G,\Phi) = s \Rightarrow \{\rho(e)\}(H_G) = s$$

By Theorem 4.54 and Lemma 4.59 we can find $(\sigma_1,\tau_1),\ldots,(\sigma_k,\tau_k)$ such that H_F is constant on each $B_{\sigma_i} \times \{\tau_i\}$ $(i = 1,\ldots,k)$, $\alpha_{\tau_i}(\sigma_i) > 0$ $(i = 1,\ldots,k)$ and whenever H_G agrees with H_F on $B_{\sigma_1} \times \{\tau_1\} \cup \ldots \cup B_{\sigma_k} \times \{\tau_k\}$, if $\{\rho(e)\}(H_G)\downarrow$ then $\{\rho(e)\}(H_G) = s$.

By definition of α_τ, each $\sigma_i \in A_{F,\tau_i}$.

By the proof of Lemma 4.58 a we see that if $\{\pi_1,\ldots,\pi_{k_i}\} = C_{\tau_i,\sigma_i}$ and if G agrees with F on $B_{\pi_1} \cup \ldots \cup B_{\pi_{k_i}}$ then H_G agrees with H_F on $B_{\sigma_i} \times \{\tau_i\}$.

By definition of A_{F,τ_i} we see that $\pi_1,\ldots,\pi_{k_i}$ are all in A_δ. So there is a finite set $\pi_1,\ldots,\pi_m$ from A_δ such that

> If F coincides with G on $B_{\pi_1} \cup \ldots \cup B_{\pi_m}$ and $\{\rho(e)\}(H_G)\downarrow$ then $\{\rho(e)\}(H_G) = \{\rho(e)\}(H_F)$

But by Lemma 4.56 we can find a G satisfying the premise and thus the conclusion above, but with $\Gamma(G) \neq \Gamma(F)$.

So if $\{e\}(G,\Phi)\downarrow$ then $\{e\}(G,\Phi) = \{\rho(e)\}(H_G) = \{\rho(e)\}(H_F) = \{e\}(F,\Phi)$

but $\Gamma(G) \neq \Gamma(F)$.

This ends the proof of Lemma 4.60.

□

We are now ready to prove the main result of this section, due to Martin Hyland.

Theorem 4.61

   Γ is not computable in the fan-functional ϕ .

Proof: Assume that Γ is computable in ϕ and let e be an index such that for all $F \in Ct(2)$

$$\Gamma(F) = \{e\}(F, \phi)$$

Fix $F \in Ct(2)$.   By Lemma 4.60 there is a $G \in Ct(2)$ such that $\{e\}(F, \phi) = \{e\}(G, \phi)$ but $\Gamma(G) \neq \Gamma(F)$, a contradiction.

□

Remark 4.62

   The proof of Theorem 4.61 can easily be relativized to any $f \in Tp(1)$. So we have indeed the stronger result (Hyland):

Theorem 4.63

   Γ is non-obtainable modulo ϕ .

□

5. THE COMPUTABLE STRUCTURE ON Ct(k)

5.1 A dense set

In chapter 3 we used the numbertheoretic relation $\text{Con}(k,\sigma,\tau)$ to prove some results about the topology T_k. If for each nonempty B_σ^k we select one $\psi_\sigma^k \in B_\sigma^k$, it is implicit in the arguments of chapter 3 that we then get a countable dense subset if $\text{Ct}(k)$. When we want to look at more effective arguments concerning computations it will be an advantage to have better control over the basic topological tools. This we will obtain by showing that $\text{Con}(k,\sigma,\tau)$ is primitive recursive and as a part of that argument find a primitive recursive dense countable subset of $\text{Ct}(k)$.

We are going to give an alternative characterization of $\text{Con}(k,\sigma,\tau)$. For $k = 1$ this characterization is trivial, we have

Lemma 5.1

$\text{Con}(1,\sigma,\tau) \leftrightarrow \sigma$ is a subsequence of τ or τ is a subsequence of σ.

Proof: If σ is a subsequence of τ we have that $B_\tau^1 \subseteq B_\sigma^1$ so $B_\sigma^1 \cap B_\tau^1 = B_\tau^1 \neq \emptyset$. If τ is a subsequence of σ the same argument works.

If neither hold there is an $n < \min\{\text{lh}(\sigma),\text{lh}(\tau)\}$ such that $\sigma(n) \neq \tau(n)$, in which case $B_\sigma \cap B_\tau = \emptyset$. $\square$

Definition 5.2

Let $\text{Con}(1,\sigma,\tau)$ hold. Let $\text{Ext}^1(t,\sigma,\tau)$ be the function f defined by

$$f(n) = \begin{cases} \sigma(n) & \text{if } n < \text{lh}(\sigma) \\ \tau(n) & \text{if } n < \text{lh}(\tau) \\ t & \text{if } n \geq \max\{\text{lh}(\sigma),\text{lh}(\tau)\} \end{cases}$$

Remark 5.3

$\text{Ext}^1(t,\sigma,\tau)$ is the extension of σ,τ by t, i.e. when neither σ nor τ carries any information about the value $f(n)$ we let $f(n) = t$.

We clearly have

Lemma 5.4

$\text{Con}(1,-,-)$ and Ext^1 are primitive recursive.

For $k = 2$ the situation is slightly more complicated. Let σ, τ be two finite sequences. We want to find out if $B^2_\sigma \cap B^2_\tau \neq \emptyset$. Let us suppose that $F \in B^2_\sigma \cap B^2_\tau$. Now if there is f, n, m such that $\sigma(\bar{f}(n)) > 0$, $\tau(\bar{f}(m)) > 0$ but $\sigma(\bar{f}(n)) \neq \tau(\bar{f}(m))$, then $F(f) = \sigma(\bar{f}(n))-1$ and $F(f) = \tau(\bar{f}(m))-1$ which clearly is impossible. In the next lemma we will see that the necessary conditions we may read out of the considerations above also will be sufficient.

Lemma 5.5

Con$(2,\sigma,\tau)$ if and only if

i $\forall \pi_1, \pi_2 < \mathrm{lh}(\sigma)$ (If $\pi_1 \prec \pi_2$ and $\sigma(\pi_1) > 0$ then $\sigma(\pi_1) = \sigma(\pi_2)$)

ii $\forall \pi_1, \pi_2 < \mathrm{lh}(\tau)$ (If $\pi_1 \prec \pi_2$ and $\tau(\pi_1) > 0$ then $\tau(\pi_1) = \tau(\pi_2)$)

iii $\forall \pi_1 < \mathrm{lh}(\sigma) \; \forall \pi_2 < \mathrm{lh}(\tau)$ (If $\sigma(\pi_1) > 0 \wedge \tau(\pi_2) > 0 \wedge \mathrm{Con}(1,\pi_1,\pi_2)$

$$\text{then } \sigma(\pi_1) = \tau(\pi_2))$$

Proof: By the considerations above we clearly have Con$(2,\sigma,\tau) \rightarrow$ **i**, **ii** and **iii**.

Now assume that **i**, **ii** and **iii** are satisfied. Let

$$\mathrm{Ext}^2(t,\sigma,\tau)(f) = \begin{cases} \sigma(\bar{f}(n))-1 & \text{if for some } n < \mathrm{lh}(\sigma) \text{ we have} \\ & \sigma(\bar{f}(n)) > 0 \\ \tau(\bar{f}(n))-1 & \text{if for some } n < \mathrm{lh}(\tau) \text{ we have} \\ & \tau(\bar{f}(n)) > 0 \\ t & \text{otherwise} \end{cases}$$

$\mathrm{Ext}^2(t,\sigma,\tau)$ is primitive recursive and symmetric in σ and τ, i.e. $\mathrm{Ext}^2(t,\sigma,\tau) = \mathrm{Ext}^2(t,\tau,\sigma)$. By **i**, **ii** and **iii** we easily see that $\mathrm{Ext}^2(t,\sigma,\tau)$ is well-defined.

If $\sigma(\pi) > 0$ then $\mathrm{Ext}^2(t,\sigma,\tau)$ is constant $\sigma(\pi)-1$ on B^1_π so $\mathrm{Ext}^2(t,\sigma,\tau) \in B^2_\sigma$. By the same argument $\mathrm{Ext}^2(t,\sigma,\tau) \in B^2_\tau$. ∎

Corollary 5.6

Con$(2,-,-)$ and Ext^2 are primitive recursive.

Proof: Directly by Lemma 5.5 and the proof of Lemma 5.5. ∎

Lemma 5.4 and Corollary 5.6 give the first two steps in an inductive proof of the main theorem.

Theorem 5.7

Let $k \geq 1$.

<u>a</u>   Con is a primitive recursive relation.

<u>b</u>   If $Con(k,\sigma,\tau)$ holds there is for each t an element $Ext^k(t,\sigma,\tau) \in B^k_\sigma \cap B^k_\tau$ primitive recursive uniformly in k,t,σ,τ and such that

$$Ext^k(t,\sigma,\tau) = Ext^k(t,\tau,\sigma)$$

<u>Proof</u>: We have defined $Con(1,-,-)$ and $Con(2,-,-)$. By induction on k we will show that the following inductive characterization of $Con(k,-,-)$ will work: $Con(k,\sigma,\tau)$ if and only if

<u>i</u> $\forall \pi_1,\pi_2 < lh(\sigma)$ (if π_2 extends π_1 and $\sigma(\pi_1) > 0$ then $\sigma(\pi_1) = \sigma(\pi_2)$)

<u>ii</u> $\forall \pi_1,\pi_2 < lh(\tau)$ (if π_2 extends π_1 and $\tau(\pi_1) > 0$ then $\tau(\pi_1) = \tau(\pi_2)$)

<u>iii</u> $\forall \pi_1,\pi_2 \ \forall \sigma_1,\sigma_2 \in \{\sigma,\tau\} \ (\sigma_1(\pi_1) > 0 \ \wedge \ \sigma_2(\pi_2) > 0 \ \wedge \ Con(k-1,\pi_1,\pi_2)$

$$\rightarrow \ \sigma_1(\pi_1) = \sigma_2(\pi_2))$$

*   We see that for $k = 2$ this characterization is identical with the one given in lemma 5.5, so this theorem is proved for $k = 1$ and $k = 2$.

** $Con(k,\sigma,\tau) \rightarrow$ <u>i</u>, <u>ii</u> and <u>iii</u> is trivial. <u>iii</u> just say that if $\phi \in B^k_\sigma \cap B^k_\tau$, $\varphi \in B^{k-1}_{\pi_1} \cap B^{k-1}_{\pi_2}$ then σ and τ cannot contain information forcing ϕ to take two different values at φ . An alternative but equivalent formulation of <u>iii</u> is

$$\forall \pi_1,\pi_2 \ \forall t_1,t_2 \ (Con(k-1,\pi_1,\pi_2) \wedge (\tau(\pi_1)=t_1+1 \vee \sigma(\pi_1)=t_1+1)$$

$$\wedge (\tau(\pi_2)=t_2+1 \vee \sigma(\pi_2)=t_2+1) \rightarrow t_1=t_2)$$

Now, let $k > 2$ and assume that the characterization is correct for all $k_1 < k$. Assume that (k,σ,τ) satisfies <u>i</u> - <u>iii</u> . We say that σ makes $\phi(\varphi)$ take the value s if for some π we have $\sigma(\pi) = s+1$ and $\varphi \in B^{k-1}_\pi$.

We will first define $Ext^k(t,\sigma,\tau)$. The idea is that if $\varphi \in Ct(k-1)$ and σ or τ makes $\phi(\varphi)$ take the value s then $Ext^k(t,\sigma,\tau)(\varphi) = s$, otherwise we would like $Ext^k(t,\sigma,\tau)(\varphi)$ to take value t . Unfortunately we may not computably decide if σ or τ make $\phi(\varphi)$ take any value at all, in contrast with the situation when $k = 2$. This is so because B^{k-1}_π is closed but in general not open and any computable set is both closed and open. But if σ or τ make

$\Phi(\varphi)$ take value s we must have that $Ext^k(t,\sigma,\tau)(\varphi) = s$ since otherwise the functional would not be in $B^k_\sigma \cap B^k_\tau$. So if $\sigma(\pi) = s+1$ (or $\tau(\pi) = s+1$) we must let $Ext^k(t,\sigma,\tau) = s$ for more φ than those in B^{k-1}_π.

Now if $\varphi \in B^{k-1}_\pi$ we can show this by computing φ on the primitive recursive functionals of type k-2 (this will be a part of the argument), so if say $\sigma(\pi_1) = s_1$, $\tau(\pi_2) = s_2 \neq s_1$ and $s_1, s_2 > 0$ then we know from <u>iii</u> that $\neg Con(k-1,\pi_1,\pi_2)$ so we may effectively find $i \in \{1,2\}$ such that $\varphi \in B^{k-1}_{\pi_i}$. This idea will be used in the formal construction of $Ext^k(t,\sigma,\tau)$.

Define

$$
Ext^k(t,\sigma,\tau)(\varphi) = \begin{cases} s & \text{if for some sequence } \delta \quad Con(k-1,\delta,\delta), \\ & (\sigma(\delta) = s+1 \text{ or } \tau(\delta) = s+1) \text{ and for all} \\ & \pi_1,\pi_2, \text{ if } Con(k-2,\pi_1,\pi_2), \ \delta(\pi_1) > 0 \text{ and} \\ & \pi_2 \leq \max\{lh(\sigma),lh(\tau)\} \text{ then} \\ & \varphi(Ext^{k-2}(0,\pi_1,\pi_2)) = \delta(\pi_1)-1 \\ \\ t & \text{otherwise} \end{cases}
$$

We must first show that $Ext^k(t,\sigma,\tau)$ is well-defined. Since the second instruction is 'otherwise' no conflicting values can arise from that. So assume that

$$Ext^k(t,\sigma,\tau)(\varphi) = s_1$$
$$Ext^k(t,\sigma,\tau)(\varphi) = s_2$$

by the first instruction.

Choose δ_1 and δ_2 demonstrating these facts. Assume $s_1 \neq s_2$.

<u>Claim</u>
$$\neg Con(k-1,\delta_1,\delta_2).$$

<u>Proof</u>: By assumption (k,σ,τ) satisfies <u>iii</u>. We do also have

$$(\sigma(\delta_1) = s_1+1 \text{ or } \tau(\delta_1) = s_1+1) \text{ and } (\sigma(\delta_2) = s_2+1 \text{ or } \tau(\delta_2) = s_2+1)$$

If we also could have $Con(k-1,\delta_1,\delta_2)$ we see easiest, from the alternative version of <u>iii</u>, that $s_1 = s_2$.
$$\square \text{Claim}$$

Now we do have $Con(k-1,\delta_1,\delta_1)$ and $Con(k-1,\delta_2,\delta_2)$ and the characterization holds for k-1. It follows from the claim that there must be π_1, π_2 such that $Con(k-2,\pi_1,\pi_2)$ with $\delta_1(\pi_1) > 0$, $\delta_2(\pi_2) > 0$ but $\delta_1(\pi_1) \neq \delta_2(\pi_2)$ establishing $\neg Con(k-1,\delta_1,\delta_2)$. Now

$\pi_2 \leq \text{lh}(\tau)$, so by the definition of $\text{Ext}^k(t,\sigma,\tau)(\varphi)$ we have

$$\varphi(\text{Ext}^{k-2}(0,\pi_1,\pi_2)) = \delta_1(\pi_1)-1$$

By a symmetric argument we have

$$\varphi(\text{Ext}^{k-2}(0,\pi_2,\pi_1)) = \delta_2(\pi_2)-1$$

But by the induction hypothesis we have

$$\text{Ext}^{k-2}(0,\pi_1,\pi_2) = \text{Ext}^{k-2}(0,\pi_2,\pi_1)$$

while $\delta_1(\pi_1) \neq \delta_2(\pi_2)$ by choice of π_1 and π_2 .
This is a contradiction so $s_1 = s_2$. It follows that $\text{Ext}^k(t,\sigma,\tau)$ is well-defined.

We see that the definition of $\text{Ext}^k(t,\sigma,\tau)$ is symmetric in σ and τ . We will show that $\text{Ext}^k(t,\sigma,\tau)$ is in B_σ^k and by symmetry it follows that it is in $B_\sigma^k \cap B_\tau^k$.

In order to show that $\text{Ext}^k(t,\sigma,\tau) \in B_\sigma^k$ it is sufficient to show that if $\sigma(\delta) > 0$ and $\varphi \in B_\delta^{k-1}$ then $\text{Ext}^k(t,\sigma,\delta)(\varphi) = \sigma(\delta)-1$. So let $\sigma(\delta) > 0$, $\varphi \in B_\delta^{k-1}$. Then for all π_1, π_2 such that $\text{Con}(k-2,\pi_1,\pi_2)$ and $\delta(\pi_1) > 0$ we have that $\text{Ext}^{k-2}(0,\pi_1,\pi_2) \in B_{\pi_1}^{k-2}$ and φ is constant $\delta(\pi_1) - 1$ on $B_{\pi_1}^{k-2}$ so $\varphi(\text{Ext}^{k-2}(0,\pi_1,\pi_2)) = \delta(\pi_1)-1$.

But then, by definition of $\text{Ext}^k(t,\sigma,\tau)$ we see that $\text{Ext}^k(t,\sigma,\tau) = \sigma(\delta)-1$.

It follows that $\text{Ext}^k(t,\sigma,\tau) \in B_\sigma^k$.

□

Remark 5.8

a   This construction has its origin in Kleene [23]. The extensions to t are slightly more general.

b   Clearly the set of the various $\text{Ext}^k(t,\sigma,\tau)$ will be a dense subset of $Ct(k)$ in the topological sense.

Definition 5.9

Let $\underline{\text{Con}(k,\sigma)}$ mean $\text{Con}(k,\sigma,\sigma)$. Let $\underline{\text{Ext}^k(t,\sigma)} = \text{Ext}^k(t,\sigma,\sigma)$.

In Theorem 5.7 we showed that the relation $B_\sigma^k \cap B_\tau^k \neq \emptyset$ is primitive recursive. Now we will show that the relation $B_\sigma^k \subseteq B_\tau^k$ also is primitive recursive.

Lemma 5.10

__a__ $B^1_\sigma \subseteq B^1_\tau$ if and only if σ is an extension of τ

__b__ $(k \geq 2)$ $B^k_\sigma \subseteq B^k_\tau$ if and only if

$$\forall \delta, s(Con(k-1, \delta) \wedge \tau(\delta) = s+1$$
$$\rightarrow \exists \pi (Con(k-1, \pi) \wedge \sigma(\pi) = s+1 \wedge B^{k-1}_\delta \subseteq B^{k-1}_\pi))$$

__c__ $(k \geq 1)$ If $B^k_\sigma \subseteq B^k_{\tau_1} \cup \ldots \cup B^k_{\tau_n}$ then there is an $i \leq n$ such that $B^k_\sigma \subseteq B^k_{\tau_i}$.

Proof: The proof is by induction on k. For $k = 1$ __a__ and __c__ are trivial, so assume $k > 1$.

__b__   The if-case is trivial so assume that $B^k_\sigma \subseteq B^k_\tau$ and let $\tau(\sigma) = s+1$. By __c__ for $k-1$ it is sufficient to show that

$$B^{k-1}_\delta \subseteq \{B^{k-1}_\pi; \sigma(\pi) = s+1\}$$

If this is not the case, let $\varphi \in B^{k-1}_\delta$ but $\varphi \notin \cup \{B^{k-1}_\pi; \sigma(\pi) = s+1\}$. Then it is easy to find $\psi \in B^k_\sigma$ such that $\psi(\varphi) \neq s$. But $\psi \in B^k_\tau$ since $B^k_\sigma \subseteq B^k_\tau$. Then ψ is constant s on B^{k-1}_δ, contradiction.

__c__   Assume that $B^k_\sigma \subseteq B^k_{\tau_1} \cup \ldots \cup B^k_{\tau_n}$, but that for no $i \leq n$ we have $B^k_\sigma \subseteq B^k_{\tau_i}$.

W.l.o.g. we may assume $Con(k, \sigma, \tau_i)$ for all $i \leq n$. If $\tau_i(\delta) > 0$ and $B^{k-1}_\delta \subseteq \cup \{B^{k-1}_\pi; \sigma(\pi) > 0\}$ there is by __c__, $k-1$ a π such that $B^{k-1}_\delta \subseteq B^{k-1}_\pi$ and $\sigma(\pi) > 0$. Since $Con(k, \sigma, \tau_i)$ we will have $\sigma(\pi) = \tau_i(\delta)$. So for each i there are δ_i, s_i such that $\tau_i(\delta_i) = s_i + 1$, and

$$B^{k-1}_{\tau_i} \nsubseteq \cup \{B^{k-1}_\pi; \sigma(\pi) > 0\}$$

(by __b__, since $B^k_\sigma \nsubseteq B^k_{\tau_i}$).

Pick $\varphi_i \in B^{k-1}_{\tau_i} \smallsetminus \cup \{B^{k-1}_\pi; \sigma(\pi) > 0\}$. Let $s > \max\{s_i; i \leq n\}$. There will be a $\psi \in B^k_\sigma$ such that $\psi(\varphi_i) = s$ for all $i \leq n$. Then $\psi \notin \underset{i \leq n}{\cup} B^k_{\tau_i}$. ☐

Corollary 5.11

__a__   The relation $'B^k_\sigma \subseteq B^k_\tau'$ is primitive recursive.

__b__   The relation $'B^k_\sigma \subseteq B^k_{\tau_1} \cup \ldots \cup B^k_{\tau_n}'$ is primitive recursive.

<u>c</u>   If $B_\sigma^k \not\subseteq B_{\tau_1}^k \cup \ldots \cup B_{\tau_n}^k$ we may find σ' extending σ such that

$$\text{Con}(k,\sigma') \wedge B_{\sigma'}^k \cap \bigcup_{i \leq n} B_{\tau_i}^k = \emptyset$$

<u>d</u>   If B_σ^k contains just one element then this element is a constant.

<u>Proof:</u> <u>a</u> and <u>b</u> are trivial from Lemma 5.10. To prove <u>c</u> we use induction on k and effectivize the proof of <u>c</u> in Lemma 5.10.
For $k = 1$ <u>c</u> is trivial, so assume $k > 1$. For each $i \leq n$ find δ_i
such that $\tau_i(\delta_i) > 0 \wedge B_{\delta_i}^{k-1} \not\subseteq \cup \{B_\pi^{k-1}; \ \sigma(\pi) > 0\}$. By the induction hypothesis we find δ_i' extending δ_i such that

$$B_{\delta_i'}^{k-1} \cap \cup\{B_\pi^{k-1}; \ \sigma(\pi) > 0\} = \emptyset$$

W.l.o.g. we may assume that $\delta_i' > \text{lh}(\sigma)$.
Let $s = 1 + \max\{\tau_i(\delta_i); \ i \leq n\}$. Then there is an extension σ' of σ
such that $\sigma'(\delta_i') = s+1$. But then

$$B_{\sigma'}^k \cap \bigcup_{i \leq n} B_{\tau_i}^k = \emptyset$$

This proves <u>c</u>.

<u>d</u>   If $k = 1$ B_σ^k will never contain just one element so we may
assume $k > 1$.

If B_σ^k contains just ψ this must mean that

$$\text{Ct}(k-1) = \cup\{B_\tau^{k-1}; \ \sigma(\tau) > 0\}$$

By Lemma 5.10 <u>c</u> there is a τ such that $\sigma(\tau) > 0$ and $\text{Ct}(k-1) \subseteq B_\tau^{k-1}$.
Then ψ is constant $\sigma(\tau) - 1$ on B_τ^{k-1} so ψ is constant $\sigma(\tau) - 1$.
                                                                                          □

<u>Lemma 5.12</u>
    The relation

$$\text{Ext}^k(t,\sigma,\tau) \in B_\pi^k$$

is primitive recursive.

<u>Proof:</u> We will use induction on k with a double induction basis.

<u>k = 1</u> $\text{Ext}^1(t,\sigma,\tau) \in B_\pi^1 \leftrightarrow \text{Ext}^1(t,\sigma,\tau)$ is an extension of π

<u>k = 2</u> $\text{Ext}^2(t,\sigma,\tau) \in B_\pi^2 \leftrightarrow \forall \delta, s \ (\pi(\delta) = s+1 \to \text{Ext}^2(t,\sigma,\tau)$ is constant s
                                                                                      on B_δ^1)

So it is sufficient to describe the relation

$$\text{Ext}^2(t,\sigma,\tau) \quad \text{is constant} \quad s \quad \text{on} \quad B_\delta^1$$

There are two cases:

__t $\ne$ s__ Then $\text{Ext}^2(t,\sigma,\tau)$ is constant s on B_δ^1 if and only if

$$\exists \delta' \!<\! \delta \ (\sigma(\delta') = s+1 \vee \tau(\delta') = s+1)$$

__t = s__ Then $\text{Ext}^2(t,\sigma,\tau)$ is constant s on B_δ^1 if and only if

$$\forall \delta',s'((\sigma(\delta') = s'+1 \vee \tau(\delta') = s'+1) \wedge \text{Con}(1,\delta,\delta') \rightarrow s' = s) \ .$$

__k > 2__ As for $k = 2$ it is sufficient to describe the relation

$$\text{Ext}^k(t,\sigma,\tau) \quad \text{is constant} \quad s \quad \text{on} \quad B_\delta^{k-1}$$

We split the argument into two cases.

__t $\ne$ s__ $\text{Ext}^k(t,\sigma,\tau)$ is constant s on B_δ^{k-1}

$$\leftrightarrow \ \forall \varphi \in B_\delta^{k-1} \text{Ext}^k(t,\sigma,\tau) = s$$

$$\leftrightarrow \ \forall \varphi \in B_\delta^{k-1} \ \exists \delta_1 \ ((\sigma(\delta_1) = s+1 \vee \tau(\delta_1) = s+1)$$

$$\wedge \forall \pi_1,\pi_2 (\delta_1(\pi_1) > 0 \wedge \pi_2 \leq \max\{\text{lh}(\sigma),\text{lh}(\tau)\}$$

$$\wedge \text{Con}(k-2,\pi_1,\pi_2) \rightarrow \varphi(\text{Ext}^2(0,\pi_1,\pi_2)) = \delta_1(\pi_1)-1))$$

$$\leftrightarrow \ \exists \delta_1 ((\sigma(\delta_1) = s+1 \vee \tau(\delta_1) = s+1)$$

$$\wedge \forall \pi_1,\pi_2 (\delta_1(\pi_1) > 0 \wedge \pi_2 \leq \max\{\text{lh}(\sigma),\text{lh}(\tau)\}$$

$$\wedge \text{Con}(k-2,\pi_1,\pi_2) \rightarrow \exists \delta'(\text{Ext}^{k-2}(0,\pi_1,\pi_2) \in B_{\delta'}^{k-2}$$

$$\wedge \delta(\delta') = \delta_1(\pi_1))))$$

All implications except the last $\rightarrow$ are trivial.
So assume $\forall \varphi \in B_\delta^{k-1} \ \text{Ext}^k(t,\sigma,\tau)(\varphi) = s$.
There will be a $\varphi \in B_\delta^{k-1}$ such that if

$$\pi_1,\pi_2 \leq \max\{\text{lh}(\sigma),\text{lh}(\tau)\} \ , \ \text{Con}(k-2,\pi_1,\pi_2) \ , \ \text{but}$$

$$\text{Ext}^2(0,\pi_1,\pi_2) \notin B_{\delta'}^{k-2} \ \text{for any} \ \delta' \ \text{such that} \ \delta(\delta') > 0$$

then

$$\varphi(\text{Ext}^{k-2}(0,\pi_1,\pi_2)) > \max\{\sigma,\tau\} \ .$$

Choose δ_1 such that

$$(\sigma(\delta_1) = s+1 \vee \tau(\delta_1) = s+1) \wedge \forall \pi_1,\pi_2(\delta_1(\pi_1) > 0 \wedge \pi_2 \leq \max\{\text{lh}(\sigma),\text{lh}(\tau)\}$$

$$\wedge \text{Con}(k-2,\pi_1,\pi_2) \rightarrow \varphi(\text{Ext}^2(0,\pi_1,\pi_2)) = \delta_1(\pi_1)-1)$$

Choose π_1, π_2 according to the premise. Since $\varphi(\text{Ext}^2(0,\pi_1,\pi_2)) < \max\{\sigma,\tau\}$ there is a δ' such that $\delta(\delta') > 0$ and $\text{Ext}^2(0,\pi_1,\pi_2) \in B_\delta^{k-2}$. Then we must also have $\delta(\delta') = \varphi(\text{Ext}^2(0,\pi_1,\pi_2))+1 = \delta_1(\pi_1)$.

<u>t = s</u> Then

$\text{Ext}^k(t,\sigma,\tau)$ is not constant s on B_δ^{k-1}

$\leftrightarrow \exists\varphi \in B_\delta^{k-1}(\text{Ext}^k(t,\sigma,\tau)(\varphi) \neq s)$

$\leftrightarrow \exists\varphi \in B_\delta^{k-1} \exists s' \neq s \; \exists\delta_1((\sigma(\delta_1) = s'+1 \vee \tau(\delta_1) = s'+1)$

$\qquad \wedge \forall\pi_1,\pi_2(\delta_1(\pi_1) > 0 \wedge \pi_2 \leq \max\{\text{lh}(\sigma),\text{lh}(\tau)\}$

$\qquad \wedge \text{Con}(k-2,\pi_1,\pi_2) \rightarrow \varphi(\text{Ext}^{k-2}(0,\pi_1,\pi_2))+1 = \delta_1(\pi_1)))$

$\leftrightarrow \exists s' \neq s \; \exists\delta_1((\sigma(\delta_1) = s'+1 \vee \tau(\delta_1) = s'+1)$

$\qquad \wedge \forall\pi_1,\pi_2(\delta_1(\pi_1) > 0 \wedge \pi_2 \leq \max\{\text{lh}(\sigma),\text{lh}(\tau)\}$

$\qquad \wedge \text{Con}(k-2,\pi_1,\pi_2) \rightarrow \forall\delta'(\delta(\delta') > 0 \wedge \text{Ext}^{k-2}(0,\pi_1,\pi_2) \in B_{\delta'}^{k-2})$

$\qquad \rightarrow \delta(\delta') = \delta_1(\pi_1)))$

Here all implications except the last $\leftarrow$ are trivial, and the last $\leftarrow$ follows by a proof similar to the one we gave in the other case.

In these descriptions we only used bounded quantifiers so the relations are clearly primitive recursive. □

5.2 The trace of a functional

In this section we let $\{\varphi_n^k\}_{k>0, n\in\omega}$ be a primitive recursive family such that

<u>i</u> $\varphi_n^k \in \text{Ct}(k)$

<u>ii</u> $\forall\sigma \; \text{Con}(k,\sigma) \rightarrow \exists n \; \varphi_n^k \in B_\sigma^k$ and we may find n primitive recursively in k,σ.

<u>iii</u> The relation $\varphi_n^k \in B_\sigma^k$ is primitive recursive.

We constructed a family with these properties in section 5.1.

In particular then $\{\varphi_n^k; n\in\omega\}$ will be a countable dense subset of $\text{Ct}(k)$.

Definition 5.13

<u>a</u>   Let $k \geq 2$, $\psi \in \text{Ct}(k)$. By the <u>trace of</u> ψ we mean

$$h_\psi(n) = \psi(\varphi_n^{k-1})$$

__b__   If $f \in Ct(1)$ we let $h_f = f$, i.e. f is its own trace.

__c__   Let $H_k = \{h_\psi ; \psi \in Ct(k)\}$.

Remark 5.14

h_ψ is uniformly primitive recursive in ψ and $\psi \mapsto h_\psi$ is one-one.

Theorem 5.15

Let $k > 1$ and let $\psi \in Ct(k)$. Then the principal associate α for ψ is recursive in the jump of h_ψ, i.e. $\alpha \in \Delta_2^0(h_\psi)$

__Proof:__ The theorem follows trivially from the following

Claim

If $Con(k-1,\sigma)$ we have

$$\alpha(\sigma) = \begin{cases} s+1 & \text{if} \quad \forall n \; \varphi_n^{k-1} \in B_\sigma^{k-1} \rightarrow h_\psi(n) = s \\ 0 & \text{if} \quad \exists n_1, n_2 (\varphi_{n_1}^{k-1} \in B_\sigma^{k-1} \wedge \varphi_{n_2}^{k-1} \in B_\sigma^{k-1} \wedge h_\psi(n_1) \neq h_\psi(n_2)) \end{cases}$$

and the claim has a trivial proof.

                □

Corollary 5.16

Let $k \geq 1$, $\psi \in Ct(k)$. Then there is a function h primitive recursive in ϕ such that

$$1\text{-}sc(\psi) \subseteq \Delta_2^0(h)$$

__Proof:__ For $k = 1$ this is trivial, so assume $k > 1$.

Each element of $1\text{-}sc(\psi)$ is recursive in any associate for ψ and hence in particular in the principal associate α, which is $\Delta_2^0(h_\psi)$. But $\Delta_2^0(h_\psi)$ is closed under recursion so $1\text{-}sc(\psi) \subseteq \Delta_2^0(h_\psi)$.

                □

Remark 5.17

Though we here see that the elements of the 1-section are rather simple, we will see in Chapter 6 that the 1-sections themselves, regarded as classes of functions, may be fairly complicated.

Corollary 5.18

Let $k \geq 1$ and let $\psi \in Ct(k)$.

Then there is a sequence $\{\psi_n\}_{n\epsilon\omega}$ of primitive recursive functions uniformly primitive recursive in ψ such that $\psi = \lim_{n\to\infty} \psi_n$.

Proof: For $k = 1$ this is trivial so assume $k > 1$.

Let α be the principal associate for ψ, h_ψ the trace of ψ. By Theorem 5.15 $\alpha \in \Delta_2^0(h_\psi)$. Then there will be a sequence $\{\alpha_i\}_{i\epsilon\omega}$ primitive recursive in h_ψ such that $\alpha = \lim_{i\to\infty} \alpha_i$. W.l.o.g. we may assume that each α_i is an associate for a functional $\varphi_{n_i}^k$. By Lemma 3.20 then $\psi = \lim_{i\to\infty} \varphi_{n_i}^k$. (To see this w.l.o.g. let $j \leq i$ be maximal such that $\text{Con}(k,\bar{\alpha}_i(j))$ and pick $\varphi_{n_i}^k \in B_{\bar{\alpha}_i(j)}^k$ uniformly primitive recursive in $\bar{\alpha}_i(j)$. Replace α_i by an associate for $\varphi_{n_i}^k$ extending $\bar{\alpha}_i(j)$.)

□

Our next result is rather trivial at this stage of development, but it is of great importance for the constructive interpretation of formulas in analysis.

Theorem 5.19
There is a partial computable selection-operator for the sets computable in a given functional.

Proof: A set is computable in Ψ if its characteristic function is computable in Ψ. For each $k > 0$ we show that there is a partial computable functional $v: Ct(k) \to Ct(k-1)$ such that for $\phi \in Ct(k)$ we have that

i $v(\phi)$ is defined if and only if $\exists\varphi \in Ct(k-1)\phi(\varphi) = 0$

ii $\exists\varphi \in Ct(k-1)\phi(\varphi) = 0 \Rightarrow \phi(v(\phi)) = 0$

Now if $\exists\varphi \in Ct(k-1)\phi(\varphi) = 0$ then $E = \phi^{-1}\{0\}$ is nonempty and both closed and open, so E will contain a φ_n^{k-1}. So let

$$v(\phi) = \varphi_n^{k-1}$$

where n is minimal such that $\phi(\varphi_n^{k-1}) = 0$.

□

Remark 5.20
v as defined above is μ-recursive. Thus this theorem is valid for the more restricted μ-recursion. Kreisel [24] called Theorem 5.19 the quantifier-free axiom of choice. He used it to give a constructive interpretation of formulas in analysis.

In Lemma 2.34 we showed that $As(k+1)$ is π_k^1 for $k \geq 1$ and in 2.35 we remarked that $As(^{k+1}0)$ will be complete π_k^1. We will now establish further connections between $Ct(k+1)$ and π_k^1.

Lemma 5.21

Let $k \geq 1$. Then H_{k+1} is π_k^1.

Proof: H_{k+1} is defined in 5.13 c.

For each h we define $\alpha_h(\sigma)$ for $Con(k,\sigma)$ by

$$\alpha_h(\sigma) = \begin{cases} s+1 & \text{if } \forall n \; \varphi_n^k \in B_\sigma^k \rightarrow h(n) = s \\ \\ 0 & \text{if } \exists n_1, n_2 \, (\varphi_{n_1}^k \in B_\sigma^k \wedge \varphi_{n_2}^k \in B_\sigma^k \wedge h(n_1) \neq h(n_2)) \end{cases}$$

If $\psi \in Ct(k+1)$ and $h = h_\psi$ then α_h will be the principal associate for ψ. On the other hand, if α_h is an associate for some ψ then $h = h_\psi$ by the following argument; given n choose σ such that $\alpha_h(\sigma) > 0$ and $\varphi_n^k \in B_\sigma^k$. Since $\alpha_h(\sigma) > 0$ we must have $h(n) = \alpha_h(\sigma) - 1 = \psi(\varphi_n^k)$.

But then

$$h \in H \leftrightarrow \alpha_h \in As(k+1)$$

so H is π_k^1.

    □

The next result was originally proved in Normann [34]. The present much simpler proof was suggested by S. Dvornickov. We use Theorem 5.19 and the proof can easily be read out of Kreisel [24].

Theorem 5.22

<u>a</u> $k \geq 2$ If A is Σ_{k-1}^1 then there is a primitive recursive relation S such that

    <u>i</u> $\alpha \in A \rightarrow \forall \psi \in Ct(k) \, \exists n \, S(\alpha, \psi, n)$

    <u>ii</u> $\alpha \notin A \rightarrow \exists \psi \, (\psi$ uniformly μ-recursive in α
              and $\forall n \neg S(\alpha, \psi, n))$

<u>b</u> $k \geq 1$ If B is π_k^1 then there is a primitive recursive relation R such that

        $\alpha \in B \leftrightarrow \forall \psi \in Ct(k) \, \exists n \, R(\alpha, \psi, n)$

Proof: We will use a simultaneous induction on k. $\underline{b},k=1$ is a well-known fact.

Claim 1

Let $k \geq 2$ be fixed. Then $\underline{a} \rightarrow \underline{b}$.

Proof: If B is Π_k^1 and A is Σ_{k-1}^1 such that

$$\alpha \in B \leftrightarrow \forall \beta <\alpha, \beta> \in A$$

choose S primitive recursive such that

$$<\alpha,\beta> \in A \leftrightarrow \forall \psi \in Ct(k) \; \exists n \; S(<\alpha,\beta>,\psi,n)$$

Then

$$\alpha \in B \leftrightarrow \forall \beta \; \forall \psi \in Ct(k) \; \exists n \; S(<\alpha,\beta>,\psi,n)$$

$$\leftrightarrow \forall <\psi_1,\psi_2> \in Ct(k) \; \exists n \; S(<\alpha,P_k^1(\psi_1)>,\psi_2,n)$$

which clearly is on the desired form.

□ Claim 1

So we are left with $\underline{b},k \rightarrow \underline{a},k+1$ for $k \geq 1$.

Let A be Σ_k^1. By $\underline{b},k$ there is a primitive recursive R such that

$$\alpha \notin A \leftrightarrow \forall \psi \in Ct(k) \; \exists n \; R(\alpha,\psi,n)$$

Then

$$\alpha \notin A \leftrightarrow \exists \phi \in Ct(k+1) \; \forall \psi \in Ct(k) \; R(\alpha,\psi,\phi(\psi))$$

where we may use $\phi_\alpha(\psi) = \mu n \, R(\alpha,\psi,n)$ when $\alpha \notin A$, cfr. Theorem 5.19.

Claim 2

Given α,ϕ the following are equivalent

i $\quad \forall \psi \in Ct(k) \; R(\alpha,\psi,\phi(\psi))$

ii $\quad \forall m \; R(\alpha,\varphi_m^k,\phi(\varphi_m^k))$

Proof: $\{\psi; R(\alpha,\psi,\phi(\psi))\}$ will be closed (and open) and $\{\varphi_m^k; m \in \omega\}$ is dense.

□ Claim 2

But then we have

$$\alpha \notin A \leftrightarrow \exists \phi \in Ct(k+1) \; \forall m \; R(\alpha,\varphi_m^k,\phi(\varphi_m^k))$$

So let $S(\alpha,\phi,m) \leftrightarrow R(\alpha,\varphi_m^k,\phi(\varphi_m^k))$.

□

We will sometimes use the Theorem in the following form. The proof is implicit in the argument above.

<u>Corollary 5.23</u>

<u>a</u> <u>$k \geq 2$</u>   If A is Σ^1_{k-1} then there is a primitive recursive
relation S such that

   <u>i</u> $\alpha \in A \rightarrow \forall h \in H_k \ \exists n \, S(\bar{\alpha}(n), \bar{h}(n), n)$

   <u>ii</u> $\alpha \notin A \rightarrow \exists \varphi \in Ct(k)$ (φ is uniformly μ-recursive in α and
$\qquad\qquad\qquad\qquad \forall n \, \neg S(\bar{\alpha}(n), \bar{h}_\varphi(n), n)$

<u>b</u> <u>$k \geq 1$</u>   If B is π^1_k then there is a primitive recursive
relation R such that

$$\alpha \in B \leftrightarrow \forall h \in H_k \ \exists n \, R(\bar{\alpha}(n), \bar{h}(n), n)$$

$\square$

5.3   The complexity of Ct(k)

In this section we will use Theorem 5.22 to show that the space Ct(k+1) is genuine π^1_k . We will use the notation from section 5.2. Some of these results, e.g. Lemma 5.24 and Corollary 5.30 were known before Theorem 5.22 and Corollary 5.23, then with more direct proofs. See Hyland [21] for details.

Lemma 5.24

Let $k \geq 1$. Then As(k+1) is complete π^1_k .

<u>Proof</u>: Let A be π^1_k . By Corollary 5.23 there is a primitive recursive relation R such that

$$\alpha \in A \longleftrightarrow \forall \varphi \in Ct(k) \ \exists n \, R(\bar{\alpha}(n), \bar{h}_\varphi(n), n)$$

We say that $\bar{h}_\varphi(n)$ is definable from σ if for all $m < n$ there is a δ such that $\sigma(\delta) > 0$ and $\varphi^{k-1}_m \in B^{k-1}_\delta$.

For each α let

$$f_\alpha(\sigma) = \begin{cases} 1 & \text{if } \exists n \leq lh(\sigma)(\bar{h}_\varphi(n) \text{ is definable from } \sigma \\ & \qquad\qquad \text{and } R(\bar{\alpha}(n), \bar{h}_\varphi(n), n)) \\ 0 & \text{otherwise} \end{cases}$$

Then

$$\alpha \in A \longleftrightarrow f_\alpha \in As(k+1) \longleftrightarrow f_\alpha \in As(^{k+1}0)$$

$\square$

Remark 5.25

Here we actually showed that $As(^{k+1}0)$ is complete π_k^1. Recall that a set is $\underset{\sim}{\pi}_k^1$ if it is $\pi_k^1(\alpha)$ for some function α.

Lemma 5.26

If $\psi \in Ct(k+1)$ then $As(\psi)$ is complete $\underset{\sim}{\pi}_k^1$.

Proof: Let α be an associate for ψ. We show that $As(\psi)$ is complete $\pi_k^1(\alpha)$. It is sufficient show that $As(^{k+1}0)$ is α-recursively reducible to $As(\psi)$.

If τ is a finite sequence, let $\pi(\beta)$ be defined by

$$\pi(\beta)(\tau) = \begin{cases} t+1 & \text{if } \beta(\tau) = 1 \text{ and } \alpha(\tau) = t+1 \text{ and } Con(k+1,\bar{\beta}(\tau+1)) \\ 0 & \text{otherwise} \end{cases}$$

Then $\beta \in As(^{k+1}0)$ if and only if $\pi(\beta) \in As(\psi)$. $\square$

These two lemmas show that $As(k+1)$ is genuine π_k^1. Later it will be useful to know this for H_{k+1} as well.

Lemma 5.27

Let $k \geq 1$. Let $\alpha \in As(k)$ and let $G(\alpha) = h_{\rho_k(\alpha)}$. Then $G : As(k) \to Tp(1)$ is continuous.

Proof: The map $h(\phi) = h_\phi$ is computable and thus continuous. ρ_k is continuous and G is the composition of h and ρ_k. $\square$

Lemma 5.28

If $A \subseteq H_{k+1}$ is $\underset{\sim}{\Sigma}_k^1$ then A is a proper subset of H_{k+1}.

Proof: All π_{k+1}^1-sets B may be defined by

$$\alpha \in B \leftrightarrow \forall h \in H_{k+1} \exists n R(\bar{\alpha}(n),\bar{h}(n),n)$$

where R is primitive recursive. If $A = H_{k+1}$ we see that all π_{k+1}^1-sets would be $\underset{\sim}{\pi}_k^1$, which is impossible. $\square$

Corollary 5.29

If $A \subseteq As(k+1)$ is $\underset{\sim}{\Sigma}_k^1$ then there is a functional $\psi \in Ct(k)$ such that ψ has no associate in A.

Proof: By Lemma 5.27 G"A will be $\underset{\sim}{\Sigma}^1_k$. The corollary then follows from Lemma 5.28.

□

Corollary 5.30

a If $\pi : Ct(n) \to Ct(m)$ is continuous and onto, then $n \geq m$.

b If $Ct(n)$ and $Ct(m)$ are homeomorphic, then $n = m$.

Proof: **b** is a direct consequence of **a**. If $n < m$ then $\{h_{\pi(\varphi)} : \varphi \in Ct(n)\}$ will be $\underset{\sim}{\Sigma}^1_n$ and in particular $\underset{\sim}{\Sigma}^1_{m-1}$. But then, by Lemma 5.28 π cannot be onto.

□

We are now going to generalize Lemma 5.26.

Lemma 5.31

Let $\psi \in Ct(k+1)$ and let $A \subseteq As(\psi)$ be $\underset{\sim}{\Sigma}^1_k$. Assume B^k_σ contains more than one element. Then there is an associate α for ψ such that

$$\forall \gamma \in A \; \exists \tau \, (\sigma \prec \tau \wedge Con(k,\tau) \wedge \gamma(\tau) > 0 \wedge \alpha(\tau) = 0)$$

Proof: We use the standard notation

$$s \overset{\cdot}{-} t = \max\{0, s-t\}$$

For $\xi \in Ct(k-1)$ we let $\xi \overset{\cdot}{-} t$ be the functional

$$(\xi \overset{\cdot}{-} t)(\eta) = \xi(\eta) \overset{\cdot}{-} t \qquad (\xi \overset{\cdot}{-} t \text{ if } k = 1)$$

Let $t > \sigma$. For $\varphi \in Ct(k)$ let

$$\varphi^t(\xi) = \begin{cases} Ext^k(0,\sigma)(\xi) & \text{if } Ext^k(s,\sigma)(\xi) \text{ is independent of } s \\ \varphi(\xi \overset{\cdot}{-} t) & \text{otherwise} \end{cases}$$

Let $\psi_0(\varphi) = \psi(\varphi^t)$.

If $\xi > t$ everywhere then, by the assumption on σ, $Ext^k(s,\sigma)(\xi)$ is independent of s (we leave this detail of the proof for the reader) so $\varphi^t(\xi) = \varphi(\xi \overset{\cdot}{-} t)$.

It follows that $\varphi \leadsto \varphi^t$ is one-to one and is a topological imbedding of $Ct(k)$ into B^k_σ. We also get a continuous imbedding of $As(k)$ into $As(k) \cap B^1_\sigma$. Take any associate for ψ and restrict it to the image of this imbedding. We then get an associate for ψ_0 by pulling the first associate back, and we may get all associates for ψ_0 this way.

The lemma is now easy to prove, we leave further details for the reader.

□

5.4 On the definability of computations

In this section we will classify semicomputability in a countable functional. We will violate the historical development by first showing that all Π^1_k-sets are semicomputable in $^{k+2}0$, and then proving the converse. These results were first proved in the opposite order. For $k = 2$ Theorem 5.32 was first proved in Bergstra [2].

Theorem 5.32

Let $A \subseteq Ct(1)$ be Π^1_k , $k > 0$. Then A is semicomputable in $^{k+2}0$.

Proof: By Theorem 5.22 $\underline{b}$ there is a primitive recursive relation R such that

$$\alpha \in A \leftrightarrow \forall \varphi \in Ct(k) \; \exists n \; R(\alpha, \varphi, n)$$

Let $\Phi_\alpha(\varphi) = \mu n \; R(\alpha, \varphi, n)$. Then

$$\alpha \in A \leftrightarrow \Phi_\alpha \text{ is total} \leftrightarrow {}^{k+2}0(\Phi_\alpha) \simeq 0$$

□

Corollary 5.33

If $\Psi \in Ct(k+2)$ and A is $\Pi^1_k(h_\Psi)$ then A is semicomputable in Ψ .

Proof: A is semicomputable in $^{k+2}0, h_\Psi$ by Theorem 5.32. The corollary follows since $^{k+2}0$ and Ψ are of the same type.

□

Remark 5.34

We do not go beyond μ-recursion in these two proofs, so the results are valid for μ-recursion as well.

We recall from 2.26 the definition of an associate and from 2.33 the equivalence-relation $\simeq_k$ on the associates. Our next lemma gives a basis for a new proof of Lemma 2.33.

Lemma 5.35

Let $k \geq 1$.

<u>a</u> $\alpha \in As(k+1)$ if and only if

$$\forall \beta \in As(k) \; \exists n \, \alpha(\bar{\beta}(n)) > 0 \; \wedge \; \forall n \; Con(k+1, \bar{\alpha}(n))$$

<u>b</u>   If $\alpha, \beta \in As(k+1)$ then

$$\alpha \approx_{k+1} \beta \leftrightarrow \forall n \; Con(k+1, \bar{\alpha}(n), \bar{\beta}(n))$$

Proof: This is proved simultaneously by induction on k . If we let $k = 0$ we see that <u>b</u> is trivial. So in proving <u>a</u> we assume that <u>b</u> holds for all $k_1 < k$.

<u>a</u>   Clearly if $\alpha \in As(k+1)$ then the right hand side must hold. So assume that the right hand side holds. From $\forall n \; Con(k+1, \bar{\alpha}(n))$ it follows that if $\alpha(\sigma) > 0$ and τ is an extension of σ then $\alpha(\tau) = \alpha(\sigma)$. This was a part of our recursive characterization of Con in section 5.1.

So it is sufficient to show that if β_1 and β_2 are associates for the same functional in $Ct(k)$, $\alpha(\bar{\beta}_1(n)) > 0$ and $\alpha(\bar{\beta}_2(n)) > 0$ then $\alpha(\bar{\beta}_1(n)) = \alpha(\bar{\beta}_2(n))$. But this is a consequence of <u>b</u> with k+1 re-placed by k , and from the fact that $\forall m \; Con(k+1, \bar{\alpha}(m))$.

<u>b</u>   Again $\rightarrow$ is trivial.

Assume that $\forall n \; Con(k+1, \bar{\alpha}(n), \bar{\beta}(n))$ and that α, β are associates for Φ_α, Φ_β resp. Let $\varphi \in Ct(k)$ and let γ be an associate for φ . Choose m such that $\alpha(\bar{\gamma}(m)) > 0$ and $\beta(\bar{\gamma}(m)) > 0$ and let $n > \bar{\gamma}(m)$. By $Con(k+1, \bar{\alpha}(n), \bar{\beta}(n))$ and $Con(k, \bar{\gamma}(m))$ it follows that $\alpha(\bar{\gamma}(m)) = \beta(\bar{\gamma}(m))$ so $\Phi_\alpha(\varphi) = \Phi_\beta(\varphi)$. Since φ was arbitrary we see that $\Phi_\alpha = \Phi_\beta$, so $\alpha \approx_{k+1} \beta$.   □

Corollary 5.36

The equivalence-relation $\approx_k$ is closed in $(As(k))^2$.

Proof: Let $K = \{(\alpha, \beta); \forall n \, Con(k, \bar{\alpha}(n), \bar{\beta}(n))\}$. Then K is closed and by Lemma 5.35

$$\approx_k = (As(k))^2 \cap K$$

□

Remark 5.37

By induction on k it follows immediately from Lemma 5.35 that $As(k) \in \Pi^1_{k-1}$.

Now we will define a set of codes for the computations $\{e\}(\Phi_1,\ldots,\Phi_n) = s$.

Definition 5.38

The set C.C. of Coded Computations is the set of sequences

$\{<e,\alpha_1,\ldots,\alpha_n,s>; \alpha_1,\ldots,\alpha_n$ are associates for

$\Phi_1,\ldots,\Phi_n$ resp. and $\{e\}(\Phi_1,\ldots,\Phi_n) = s\}$

Definition 5.39

Let f be the partial recursive function constructed in Theorem 2.28.

We write

$$\{e\}_I(\alpha_1,\ldots,\alpha_n) \simeq s$$

if there is an m such that

$$f(e,\bar{\alpha}_1(m),\ldots,\bar{\alpha}_n(m)) = s+1$$

Remark 5.40

The index I stands for Intentional computation. Clearly the map $g(e,\alpha_1,\ldots,\alpha_n) \simeq \{e\}_I(\alpha_1,\ldots,\alpha_n)$ is partially recursive. As a consequence of Theorem 2.28 we see that g is well-defined.

We will now show that C.C. is inductively defined.

Lemma 5.41

The following induction in nine cases defines C.C. (We give four cases and leave the rest for the reader.)

1. If $e = <0,x,\sigma>$ and $\alpha_1,\ldots,\alpha_n$ are associates for elements of the appropriate form for σ , then $<e,x,\alpha_1,\ldots,\alpha_n,x+1> \in C.C.$

4. If $e = <4,e_1,e_2,\sigma>$, $<e_2,\alpha_1,\ldots,\alpha_n,s> \in C.C.$ and $<e_1,s,\alpha_1,\ldots,\alpha_n,t> \in C.C.$ then $<e,\alpha_1,\ldots,\alpha_n,t> \in C.C.$

8. If $e = <8,e_1,\sigma>$ and for all $\beta \in As(\sigma(0))$ there is an s such that $<e_1,\beta,\alpha_1,\ldots,\alpha_n,s> \in C.C.$ then $<e,\alpha_1,\ldots,\alpha_n,t> \in C.C.$ where $t \simeq \{e\}_I(\alpha_1,\ldots,\alpha_n)$.

9. If $e = <9,\sigma,\tau>$ and $<x,\alpha_1,\ldots,\alpha_n,s> \in C.C.$ then $<e,x,\alpha_1,\ldots,\alpha_n,\beta_1,\ldots,\beta_n,s> \in C.C.$ where we assume that all α's and β's are associates for functionals of the correct type.

Proof: To show that if $\langle e, \alpha_1, \ldots, \alpha_n, s \rangle \in C.C.$ as defined in 5.39 then $\langle e, \alpha_1, \ldots, \alpha_n, s \rangle$ is generated by this induction we use a long, uninspiring induction on the computations $\{e\}(\phi_1, \ldots, \phi_n) = s$.

The other direction follows by an induction on the definition suggested in this lemma. Again a formal proof is long and adds no new insight. As we see we have just imitated the definition of Kleene-computations using associates instead of functionals. We leave the details for the reader.

□

Remark 5.42

At first sight the inductive definition given in Lemma 5.41 looks frightfully complicated. But a closer inspection reveals that the only complicated case is case 8 where we write

$$\forall \beta(\beta \in As(\sigma(0)) \ldots)$$

so the definition is $\pi_1^1(\langle As(k) \rangle_{k \in \omega})$.

What is more important is that if $\alpha_1, \ldots, \alpha_n$ are associates for elements of type $\leq k$ then in order to establish that $\langle e, \alpha_1, \ldots, \alpha_n, s \rangle \in C.C.$ we only need to use

$$\forall \beta(\beta \in As(\sigma(0)) \ldots)$$

where $\sigma(0) \leq k-2$ (see Lemma 1.17 b). Moreover the induction is positive, we did not use the negation symbol in defining it. This gives us

Lemma 5.43

Let $k \geq 2$. Let $\phi \in Ct(k)$ and let α be the principal associate for ϕ. Let

$$A_\phi = \{\langle e, \alpha_1, \ldots, \alpha_n, s \rangle \in C.C.; \text{ for all } i, \text{ either } \alpha = \alpha_i \text{ or}$$
$$\alpha_i \in As(j) \text{ for some } j \leq k-2\}$$

Then A_ϕ is definable by a $\pi_1^1(\langle As(i) \rangle_{i \leq k-2})$-positive inductive definition relative to a function computable in ϕ.

Proof: Immediate from the considerations in Remark 5.42.

□

Using this lemma we may use results from the literature to show that A_ϕ is π_{k-2}^1 in the principal associate for ϕ. In Cenzer [5] and in Moldestad-Normann [28] the following theorem is shown.

Proposition 5.44

Let Γ be a $\pi_1^1(B)$ - positive inductive definition over $Tp(1)$. Then the closure Γ^∞ is $\pi_1^1(B)$.

□

If $k \geq 3$ we may clearly use Lemma 5.43 and this proposition to show that A_ϕ is $\pi_{k-2}^1(\alpha)$ where α is the principal associate for ϕ . We have however, not assumed intimate knowledge of the theory of inductive definitions so we will give a direct argument following J. Bergstra [1].

Theorem 5.45

Let $k \geq 3$, $\phi \in Ct(k)$. Let α be the principal associate for ϕ and let A_ϕ be defined as in Lemma 5.43.

Then A_ϕ is $\pi_{k-2}^1(\alpha)$.

Proof: We let $<e, \alpha_1, \ldots, \alpha_n, s> \in B_\phi$ if the following are satisfied:

i̲ e is an index for $\{e\}(\varphi_1, \ldots, \varphi_n)$ such that if $\varphi_i \in Tp(j)$, then $\forall n \, Con(j, \bar{\alpha}_j(n))$ and either $j \leq k-2$ or $\alpha_j = \alpha$ $(j = 1, \ldots, n)$.

ii̲ $\{e\}_I(\alpha_1, \ldots, \alpha_n) \simeq s$ (see Definition 5.39).

We let

$$O_\phi = \{<e, \alpha_1, \ldots, \alpha_n> \text{ satisfying } \underline{i} \text{ above such that } \alpha_1, \ldots, \alpha_n \text{ are real associates}\}$$

We let $<e, \alpha_1, \ldots, \alpha_n, s> \in C_\phi$ if $<e, \alpha_1, \ldots, \alpha_n, s> \in B_\phi$ and for each $i \leq n$ we have $\alpha_i = \alpha$ or $\alpha_i \in As(j)$ for the appropriate j .

Now B_ϕ is arithmetical in α and $As(j)$ is π_{j-1}^1 so C_ϕ will be $\pi_{k-3}^1(\alpha)$ (arithmetical for $k = 3$).

If $<e, \alpha_1, \ldots, \alpha_n, s> \in C.C.$ with $<e, \alpha_1, \ldots, \alpha_n> \in O_\phi$ then $<e, \alpha_1, \ldots, \alpha_n, s> \in C_\phi$. Moreover, if $<e, \alpha_1, \ldots, \alpha_n, s> \in C_\phi$ then s is uniquely determined by $e, \alpha_1, \ldots, \alpha_n$ (see ii̲ of the definition of B_ϕ). We have to select the 'correct computations' among the elements of C_ϕ .

We will define a relation R on pairs of $n+2$-tuples $<e, \alpha_1, \ldots, \alpha_n, s>$ where $<e, \alpha_1, \ldots, \alpha_n> \in O_\phi$.

On C_ϕ we will try to imitate the immediate subcomputation-relation and we want the real computations to be in the well-founded part of R . Let $<e, \alpha_1, \ldots, \alpha_n, s> \notin C_\phi$. Then

$$R(<e, \alpha_1, \ldots, \alpha_n, s>, <e, \alpha_1, \ldots, \alpha_n, s>) .$$

If $<e,\alpha_1,\ldots,\alpha_n,s> \in C_\phi$ then the definition of

$$R(-,<e,\alpha_1,\ldots,\alpha_n,s>)$$

is directly imitating the immediate sub-computation-relation in all cases except case 8: $e = <8,e_1,\sigma>$. (In these cases the immediate 'sub-computations' will be in C_ϕ as well.)

So let $e = <8,e_1,\sigma>$, $j = \sigma(0)$. Then $j \leq k$. Let $\beta \in As(j-2)$. If for some t

$$<e_1,\beta,\alpha_1,\ldots,\alpha_n,t> \in C_\phi$$

we let $R(<e_1,\beta,\alpha_1,\ldots,\alpha_n,t>, <e,\alpha_1,\ldots,\alpha_n,s>)$ for that t. If for no t $<e_1,\beta,\alpha_1,\ldots,\alpha_n,t> \in C_\phi$ then

$$R(<e_1,\beta,\alpha_1,\ldots,\alpha_n,t>, <e,\alpha_1,\ldots,\alpha_n,s>) \quad \text{for all } t.$$

R is clearly arithmetic in C_ϕ, O_ϕ and thus $\Delta^1_{k-2}(\alpha)$.

Claim

Assume that $<e,\alpha_1,\ldots,\alpha_n,s> \in C_\phi$. Then $<e,\alpha_1,\ldots,\alpha_n,s> \in C.C.$ if and only if there is no R-descending chain starting at $<e,\alpha_1,\ldots,\alpha_n,s>$ (where an R-descending chain is a sequence of tuples $\{\vec{\alpha}_i\}_{i\in\omega}$ such that for all i $R(\vec{\alpha}_{i+1},\vec{\alpha}_i)$).

Proof: If $<e,\alpha_1,\ldots,\alpha_n,s>$ codes the computation $\{e\}(\varphi_1,\ldots,\varphi_n) \simeq s$ we show by induction on the length of this computation that $<e,\alpha_1,\ldots,\alpha_n,s> \in C_\phi$ and that there is no R-descending chain starting at $<e,\alpha_1,\ldots,\alpha_n,s>$. There are no new tricks or ideas in this argument and we leave it for the reader to sort it out.

Now assume that there is no R-descending chain starting at $<e,\alpha_1,\ldots,\alpha_n,s>$. We must have that $<e,\alpha_1,\ldots,\alpha_n,s> \in C_\phi$ since otherwise $\vec{\alpha}_i = <e,\alpha_1,\ldots,\alpha_n,s>$ would make up an R-descending chain.

Moreover, by the same argument it follows that if we move downwards along R a finite number of steps from $<e,\alpha_1,\ldots,\alpha_n,s>$ we will stay in C_ϕ.

Let $\alpha_1,\ldots,\alpha_n$ be associates for $\varphi_1,\ldots,\varphi_n$ resp. We cannot have that $\{e\}(\varphi_1,\ldots,\varphi_n) = s_1 \neq s$ since then $<e,\alpha_1,\ldots,\alpha_n,s_1> \in C_\phi$ and s is uniquely determined by $<e,\alpha_1,\ldots,\alpha_n>$.

So assume that $\{e\}(\varphi_1,\ldots,\varphi_n)\uparrow$. To obtain a contradiction it is sufficient to construct an R-descending chain starting at $<e,\alpha_1,\ldots,\alpha_n,s>$. But indeed, a descending chain of subcomputations of $\{e\}(\varphi_1,\ldots,\varphi_n)$ will translate to an R-descending chain, by construction of R.

$\square$ Claim

It follows from the claim that if $<e,\alpha_1,\ldots,\alpha_n,s> \in C_\phi$ then

$$<e,\alpha_1,\ldots,\alpha_n,s> \in C.C. \quad \text{if and only if}$$

$$\forall \vec{\alpha}_i(\forall i(\vec{\alpha}_i \in C_\phi) \wedge \vec{\alpha}_0 = <e,\alpha_1,\ldots,\alpha_n,s> \Rightarrow \exists i \neg R(\vec{\alpha}_{i+1},\vec{\alpha}_i))$$

This gives a $\pi^1_{k-2}(\alpha)$-definition of A_ϕ .

□

Corollary 5.46

Let $k \geq 3$, $\phi \in Ct(k)$. Then there is a function h primitive recursive in ϕ such that

$$2\text{-en}(\phi) \subseteq \pi^1_{k-2}(h)$$

Proof: Let $h = h_\phi$ be the trace of ϕ (5.13 a). By Theorem 5.15 the principal associate α for ϕ is $\Delta^0_2(h)$. Let $A \in 2\text{-en}(\phi)$. Then there is an index e such that

$$f \in \Lambda \longleftrightarrow \{e\}(f,\phi) \simeq 0$$

Let A_ϕ be as in Lemma 5.43. Then

$$f \in A \longleftrightarrow <e,f,\alpha,0> \in A_\phi$$

But by Theorem 5.45 A_ϕ will be $\pi^1_{k-2}(\alpha)$ so A is $\pi^1_{k-2}(\alpha)$. Now h is recursive in α and $\alpha \in \Delta^0_2(h)$ so $\pi^1_{k-2}(\alpha) = \pi^1_{k-2}(h)$. It follows that $A \in \pi^1_{k-2}(h)$.

□

We have now proved the following result.

Theorem 5.47

Let $k \geq 3$, $\phi \in Ct(k)$. Let $h = h_\phi$ be the trace of ϕ (5.13 a). Let $A \subseteq Tp(1)$. Then the following are equivalent:

<u>i</u> A is μ-semirecursive in ϕ

<u>ii</u> A is semicomputable in ϕ

<u>iii</u> A is $\pi^1_{k-2}(h)$

Proof:

<u>i</u> → <u>ii</u>   is obvious

<u>ii</u> → <u>iii</u>   is Corollary 5.46

<u>iii</u> → <u>i</u>   see Corollary 5.33 and Remark 5.34

□

So for the envelopes of functionals of type ≥ 3 the μ-semicomputable sets and the Kleene-semicomputable sets are the same. This will not be the case for elements of $Ct(2)$. In Corollary 2.11 we saw that $2\text{-en}(^2 0) = \Pi_1^1$. For μ-computations we have

<u>Lemma 5.48</u>

Let $A \subseteq Tp(1)$. Then the following are equivalent.

<u>i</u> A is μ-semirecursive in $^2 0$

<u>ii</u> A is Π_2^0

<u>Proof</u>: $A \in \Pi_2^0$ let R be recursive such that

$$f \in A \leftrightarrow \forall n \, \exists m \, R(f,n,m)$$

Let $g_f(n) = \mu m \, R(f,n,m)$. Then

$$f \in A \leftrightarrow g_f \text{ is total} \leftrightarrow {}^2 0(g_f) \simeq 0$$

This shows <u>ii</u> $\rightarrow$ <u>i</u>.

On the other hand if A is μ-semirecursive in $^2 0$ then there is an index e for a primitive recursive function $\{e\}(\mu,{}^2 0,f)$ such that

$$f \in A \leftrightarrow \{e\}(\mu,{}^2 0,f) \simeq 0$$

(See Definition 4.22.)

Inductively we define the relation

$$\{e\}_n(\mu,{}^2 0,f,\vec{x}) \simeq k$$

as for the Kleene-computation $\{e\}(\mu,{}^2 0,f\ \vec{x})$ with the one change that if e is an index for $S8$ involving an application of $^2 0$

$$\{e\}(\mu,{}^2 0,f,\vec{x}) \simeq {}^2 0(\lambda y \{e_1\}(\mu,{}^2 0,f,\vec{x},y))$$

then we let

$$\{e\}_n(\mu,{}^2 0,f,\vec{x}) = 0 \quad \text{if} \quad \{e_1\}_n(\mu,{}^2 0,f,\vec{x},y)$$

is defined for all $y \leq n$.

<u>Claim</u>

<u>a</u>   If e is an index for a Kleene-computation and $\{e\}(\mu,{}^2 0,f,\vec{x}) \simeq k$ then $\{e\}_n(\mu,{}^2 0,f,\vec{x}) \simeq k$ for all n.

<u>b</u>   If e is an index for a Kleene-computation, $\{e\}_m(\mu,{}^2 0,f,\vec{x}) \simeq k$ and $n < m$ then

$$\{e\}_n(\mu,{}^20,f,\vec{x}) \simeq k$$

<u>c</u>   If e is an index for a primitive recursive Kleene-computation and

$$\forall n \, \{e\}_n(\mu,{}^20,f,\vec{x}) \simeq k$$

then

$$\{e\}(\mu,{}^20,f,\vec{x}) \simeq k$$

Proof: Both <u>a</u> and <u>b</u> have trivial proofs by induction on the length of computations.

We prove <u>c</u> by induction on e noticing that for primitive recursion the following holds

If $\{e_1\}(\mu,{}^20,\vec{y})$ is a proper subcomputation

of $\{e\}(\mu,{}^20,\vec{x})$ then $e_1 < e$.

We restrict ourselves to proving two cases, the rest of the proof is trivial.

<u>1</u> $\{e\}(\mu,{}^20,f,\vec{x}) = \mu x(\{e_1\}(\mu,{}^20,f,\vec{x},x) \simeq 0)$

Assume that for all n $\{e\}_n(\mu,{}^20,f,\vec{x}) \simeq x$. This means that

<u>i</u> $\forall y < x \, \{e_1\}_n(\mu,{}^20,f,\vec{x},y)$ is defined and $\neq 0$

<u>ii</u> $\forall n \, \{e_1\}_n(\mu,{}^20,f,\vec{x},x) = 0$

From <u>i</u> and <u>b</u> there will for each $y < x$ be a number $k_y \neq 0$ such that

$$\forall n \, \{e_1\}_n(\mu,{}^20,f,\vec{x},y) \simeq k_y$$

By the induction-hypothesis we see that for $y < x$ we have

$$\{e_1\}(\mu,{}^20,f,\vec{x},y) = k_y \neq 0$$

and

$$\{e_1\}(\mu,{}^20,f,\vec{x},x) = 0$$

so

$$\{e\}(\mu,{}^20,f,\vec{x}) = x$$

<u>2</u> $\{e\}(\mu,{}^20,f,\vec{x}) \simeq {}^20(\lambda y \{e_1\}(\mu,{}^20,f,\vec{x},y))$

Assume that $\forall n \, \{e\}_n(\mu,{}^20,f,\vec{x}) \simeq 0$. This means that for all y and for all $n > y$

$$\{e\}_n(\mu,{}^20,\vec{x},y)$$ is defined.

By $\underline{b}$ it follows that

$$\forall y \ \forall n \ \{e_1\}_n(\mu,{}^2O,f,\vec{x},y) \quad \text{is defined}$$

and the value is independent of n.

By the induction-hypothesis then

$$\forall y \ \{e_1\}(\mu,{}^2O,f,\vec{x},y) \quad \text{is defined}$$

so

$$\lambda y \ \{e_1\}(\mu,{}^2O,f,\vec{x},y) \quad \text{is total}$$

and

$$\{e\}(\mu,{}^2O,f,\vec{x}) \simeq 0$$

$$\square \text{ Claim}$$

From the claim it follows that

$$f \in A \ \leftrightarrow \ \forall n \ \{e\}_n(\mu,{}^2O,f) \simeq 0$$

where e is an index for a primitive recursive function such that $f \in A \leftrightarrow \{e\}(\mu,{}^2O,f) \simeq 0$.

Clearly the function $\{e\}_n$ is partially recursive uniformly in n so the relation

$$\{e\}_n(\mu,{}^2O,f) \simeq 0$$

is Σ_1^0.

It follows that A is Π_2^0. This establishes $\underline{i} \rightarrow \underline{ii}$.

$$\square$$

Remark 5.49

The relation

$$\text{Tot} = \{e; \ \lambda x \{e\}(x) \text{ is total}\}$$

is known to be complete Π_2^0 over ω. Thus Lemma 5.48 indicates that with μ-recursion in ${}^2O$ we can only check totality of recursive functions while Kleene-computations give the power to evaluate Π_1^0 positive inductive definitions.

5.5 Regularity of countable recursion

In section 2.3 we defined the relation

$$[e](\psi_1,\ldots,\psi_k) \simeq t$$

to mean:

Whenever $\beta_1,\ldots,\beta_k$ are associates for $\psi_1,\ldots,\psi_k$ resp. then $\{e\}(\beta_1,\ldots,\beta_k) \simeq t$.

Given e we may define the function h_e by

$$h_e(\bar{\beta}_1(n),\ldots,\bar{\beta}_k(n)) \simeq \begin{cases} t+1 & \text{if } \{e\}(\beta_1,\ldots,\beta_k) \simeq t \text{, the} \\ & \text{computation takes less than } n \text{ steps} \\ & \text{and we only use information from} \\ & \bar{\beta}_1(n),\ldots,\bar{\beta}_k(n) \\ \\ 0 & \text{otherwise} \end{cases}$$

Now the index e can be any odd index, never designed to act on associates but accidentally doing so. There may be many ways to restrict the notion of recursion such that the algorithms used really reflect that they work on initial segments of associates. We are going to suggest one such restriction, and then show that all partial recursive functionals will be covered by this restricted notion as well. This will actually improve our ability to handle recursion and contribute to justify the notion.

Definition 5.50

Let the types $\bar{t} = t_1,\ldots,t_k$ be fixed. Let e be an index operating on k functions.

We say that e is $\underline{\bar{t}\text{-operational}}$ if for all $\alpha_1,\ldots,\alpha_k,\beta_1,\ldots,\beta_k,n,m$:

If $Con(t_i,\bar{\alpha}_i(n),\bar{\beta}_i(m))$ $(i=1,\ldots,k)$ and
$h_e(\bar{\alpha}_1(n),\ldots,\bar{\alpha}_k(n)) > 0$ and $h_e(\bar{\beta}_1(m),\ldots,\bar{\beta}_k(m)) > 0$ then
$h_e(\bar{\alpha}_1(n),\ldots,\bar{\alpha}_k(n)) = h_e(\bar{\beta}_1(m),\ldots,\bar{\beta}_k(m))$.

Remark 5.51

e is $\bar{t}$-operational if $\{e\}(\alpha_1,\ldots,\alpha_k)$ and $\{e\}(\beta_1,\ldots,\beta_k)$ must be equal as long as consistent information from $\alpha_1,\ldots,\alpha_k$ and $\beta_1,\ldots,\beta_k$ is used in these computations. Clearly any index obtained from translating a computation to a recursion is operational. The recursions via operational indices are better behaved than others, as we will see in Corollary 5.53.

Theorem 5.52

Let the types $\bar{t} = t_1,\ldots,t_k$ be fixed. Let e be an index. Then there is a $\bar{t}$-operational index e_1 such that

$$\lambda \psi_1,\ldots,\psi_k[e](\psi_1,\ldots,\psi_k) = \lambda \psi_1,\ldots,\psi_k[e_1](\psi_1,\ldots,\psi_k)$$

where ψ_i varies over $Ct(k_i)$.

Proof: We assume that $\vec{t}$ contains one element $t \geq 2$. The general proof only requires more notation.

Let $A = \{(s,\tau); \{e\}_s(\tau) \text{ is defined}\}$

A has a recursive enumeration $A = \{(s_i,\tau_i); i \in \omega\}$. Let e_1 be an index for the following computation $\{e_1\}(\alpha)$:

First find x and minimal s such that

$$\{e\}_s(\bar{\alpha}(s)) = x$$

Let i_0 be such that $(s,\bar{\alpha}(s)) = (s_{i_0},\tau_{i_0})$. For each $i \leq i_0$, if $\{e\}_{s_i}(\tau_i) \neq x$ find r_i such that $\neg Con(t,\tau_i,\bar{\alpha}(r_i))$. Then let $\{e_1\}(\alpha) = x$.

Claim 1

$$\lambda\psi[e_1](\psi) \subseteq \lambda\psi[e](\psi)$$

This is trivial from the first instruction for $\{e_1\}(\alpha)$.

Claim 2

$$\lambda\psi[e](\psi) \subseteq \lambda\psi[e_1](\psi)$$

Proof: Let $\psi \in Ct(t)$, let α be an associate for ψ and assume that $[e](\psi) = x$.

Find s,i_0 as above. If $i \leq i_0$ and $\{e\}_{s_i}(\tau_i) \neq x$ then τ_i cannot be extended to an associate for ψ . By lemma 3.13 there is an r_i such that $\neg Con(t,\tau_i,\bar{\alpha}(r_i))$. When we have found all such r_i we can let $\{e_1\}(\alpha) = x$.   □ Claim 2

Claim 3

e_1 is t-operational.

Proof: Assume that $\{e_1\}(\tau) = x$, $\{e_1\}(\pi) = y$ and that $x \neq y$. Let s_1,s_2 be minimal such that

$$\{e\}_{s_1}(\bar{\tau}(s_1)) = x \qquad \{e\}_{s_2}(\bar{\pi}(s_2)) = y$$

Let i_1,i_2 be such that $(s_1,\bar{\tau}(s_1)) = (s_{i_1},\tau_{i_1})$ and $(s_2,\bar{\pi}(s_2)) = (s_{i_2},\tau_{i_2})$. W.l.o.g. we may assume that $i_1 \leq i_2$. In order to compute $\{e_1\}(\pi)$ we should then find r_i such that $\neg Con(t,\bar{\tau}(s_1),\bar{\pi}(r_i))$. So in particular

$\daleth Con(k,\tau,\pi)$, which was what we wanted to prove.

□ Claim 3

The theorem follows from the claims.   □

Corollary 5.53

Let $\psi \in Ct(k)$ be arbitrary. Then there is a function h primitive recursive in ψ such that whenever f is recursive in ψ then there is an r.e.(h)-set D recursive in ψ such that f is recursive in D .

Proof: Let $f = \lambda x[e](\psi,x)$. By Theorem 5.52 we may assume that e is operational.

From the proofs of Theorem 5.15 and Corollary 5.18 we see that there is a sequence $\{\sigma_n\}_{n \in \omega}$ of finite sequences primitive recursive in ψ such that $\lim_{n \to \infty} \sigma_n$ is the principal associate for ψ . σ_n is uniformly recursive in any associate α for ψ . We may assume that $Con(k,\sigma_n)$ and that $lh(\sigma_n) = n$. Moreover, for each associate α for ψ and each n we may assume that if $\pi < n$ then $\alpha(\pi) = 0$ or $\alpha(\pi) = \sigma_n(\pi)$.

Let h_e be as in the beginning of this section and let

$$D = \{(n,x) : \exists k > n \ h_e(\sigma_k,x) \neq h_e(\sigma_n,x)\} .$$

Then D is r.e. in the trace h_ψ of ψ .

To compute f(x) from D we find the least n such that $(n,x) \notin D$ and let $f(x) = h_e(\sigma_n,x)$. It remains to show that D is recursive in ψ .

Let (n,x) be given, α any associate for ψ . Find s such that $\{e\}(\alpha,x)$ is computable from $\bar{\alpha}(s)$. For $m > s$ we will have $Con(k,\sigma_m,\bar{\alpha}(s))$ so find the least m such that $\{e\}(\sigma_m,x)\downarrow$. Since e is operational we must have that

$$\{e\}(\sigma_m,x) = \{e\}(\bar{\alpha}(s),x) .$$

Then $(n,x) \in D \leftrightarrow \exists k(n < k \leq m \wedge h_e(\sigma_k,x) \neq h_e(\sigma_n,x)) .$ □

Remark 5.54

We say that the set of functions recursive in ψ is generated by its r.e.-degrees modulo h_ψ . This will be a phenomenon that we will come back to in connection with Kleene-computations.

6. SECTIONS

6.1 1-sections in a general type-structure

In section 5.5 we showed that countable 1-sections of continuous functionals are generated by their r.e. degrees modulo some function. In Normann-Wainer [35] it is shown that the same is valid for Kleene-computations in an arbitrary countable functional. The main method is the same, all Kleene-computations are approximated such that the modulus of convergence is computable.

Here we will give an even stronger result. In the maximal type-structure we will analyze for which functionals ϕ 1-sc(ϕ) is generated by its r.e. degrees. Our conclusion will be that this holds if and only if ^2E is not computable in ϕ . This will in particular verify a conjecture of J. Bergstra [1]; the following are equivalent:

i ^2E is computable in ϕ

ii 1-sc(ϕ) is closed under jump

In this section we will work within the maximal type-structure, but the arguments are valid for all type-structures closed under computations.

Our strategy will be as follows. In a naive way we will try to approximate all Kleene-computations by primitive recursive functions. If we succeed we may computably find the modulus of the convergence. If we fail we may utilize this failure to compute ^2E from the arguments involved. Moreover we may by a computation decide whether we succeed or fail.

Our n'th approximation to a computation $\{e\}(\vec{\psi})$ will be what we can obtain by completing n steps of this computation. To be slightly more precise we look at the computation tree for $\{e\}(\vec{\psi})$ and follow each path n steps downwards. If we reach an initial computation we put out the correct value. Otherwise we put out zero at the end of the path and then fill in the values of each computation in this tree.

In the following definition we give a precise definition $h(n,e,\vec{\psi})$ of the n'th approximation to $\{e\}(\vec{\psi})$.

Definition 6.1

Let e be a number, $\vec{\psi}$ a finite sequence of functionals. By the __n'th approximation__ to $\{e\}(\vec{\psi})$ we mean $h(n,e,\vec{\psi})$ defined inductively as follows

i $h(0,e,\vec{\psi}) = 0$

ii $h(n+1,e,\vec{\psi})$ is defined by ten cases, the nine first corresponding to the indices for S1 - S9.

1-3,7: If $\{e\}(\vec{\psi})$ is an initial computation we let

$$h(n+1,e,\vec{\psi}) = \{e\}(\vec{\psi})$$

4   If $e = <4,e_1,e_2,\sigma>$ we let

$$h(n+1,e,\vec{\psi}) = h(n,e_1,h(n,e_2,\vec{\psi}),\vec{\psi})$$

5-6   are analogous to 4

8   If $e = <8,e_1,\sigma>$, let

$$h(n+1,e,\vec{\psi}) = \psi_1(\lambda\varphi h(n,e_1,\varphi,\vec{\psi}))$$

9   If $e = <9,\sigma,\tau>$ let

$$h(n+1,e,x,\vec{\psi},\vec{\phi}) = h(n,x,\vec{\psi})$$

10   Otherwise, let $h(n+1,e,\vec{\psi}) = 0$

In 1-9 it is understood that the types in $\vec{\psi}$ correspond to σ .

Remark 6.2

    h as defined in 6.1 is clearly primitive recursive so $h(n,e,\vec{\psi})$ is defined for all $n,e,\vec{\psi}$.

Theorem 6.3

    Let h be as in Definition 6.1. Then there are partial computable functions E,M such that

i $\{e\}(\vec{\psi})\!\downarrow \leftrightarrow E(e,\vec{\psi})\!\downarrow$

ii If $E(e,\vec{\psi}) = 0$ then $M(e,\vec{\psi})\!\downarrow$,

$$\{e\}(\vec{\psi}) = \lim_{n\to\infty} h(n,e,\vec{\psi}) \text{ and}$$
$$\forall n \geq M(e,\vec{\psi})(\{e\}(\vec{\psi}) = h(n,e,\vec{\psi}))$$

iii If $E(e,\vec{\psi}) > 0$ then ^2E is computable in $\vec{\psi}$ via the index $E(e,\vec{\psi})$

Proof: Formally we will produce E_σ and M_σ uniformly recursive in σ working for $\vec{\psi} \in Tp(\sigma(0))\times\ldots\times Tp(\sigma(n-1))$ where $n = lh(\sigma)$. But in order to avoid notational difficulties we use E,M as a common name for all these.

E and M are constructed by the recursion theorem, and as usual we will give the construction by induction on the length of the computation $\{e\}(\vec{\psi})$ and at the same time show that it works.

Clearly $E(e,\vec{\psi}) = 0$ means that the approximation works and then $M(e,\vec{\psi})$ gives a modulus for the convergence.

We will now construct E and M .

<u>1-3,7</u> If $\{e\}(\vec{\psi})$ is an initial computation we let $E(e,\vec{\psi}) = 0$ and $M(e,\vec{\psi}) = 1$

<u>4</u>   If $\{e\}(\vec{\psi}) = \{e_1\}(\{e_2\}(\vec{\psi}),\vec{\psi})$ we regard two cases:

<u>i</u> $E(e_2,\vec{\psi}) = 0$ and $E(e_1,\{e_2\}(\vec{\psi}),\vec{\psi}) = 0$. Then let $E(e,\vec{\psi}) = 0$ and $M(e,\vec{\psi}) = 1 + \max\{M(e_1,\vec{\psi}),M(e_2,\{e_1\}(\vec{\psi}),\vec{\psi})\}$

<u>ii</u> $E(e_2,\vec{\psi}) > 0$ or $E(e_1,\{e_2\}(\vec{\psi}),\vec{\psi}) > 0$
If $E(e_1,\{e_2\}(\vec{\psi}),\vec{\psi}) > 0$ then it is an index for computing ^2E from $(\{e_2\}(\vec{\psi}),\vec{\psi})$. Let $E'(e_1,\{e_2\}(\vec{\psi}),\vec{\psi}))$ be an index > 0 for computing ^2E from $\vec{\psi}$ uniformly obtained from $E(e_1,\{e_2\}(\vec{\psi}),\vec{\psi})$ and $\{e_2\}(\vec{\psi})$. Let

$$E(e,\vec{\psi}) = \max\{E(e_2,\vec{\psi}),E'(e_1,\{e_2\}(\vec{\psi}),\vec{\psi})\}$$

<u>5</u>, <u>6</u> and <u>9</u> are treated similarly and all proofs involved are trivial. The problematic part comes in dealing with case 8:

<u>8</u> $\{e\}(\vec{\psi}) = \psi_1(\lambda\varphi\{e_1\}(\varphi,\vec{\psi}))$

We now assume that the theorem holds for all computations $\{e_1\}(\varphi,\vec{\psi})$.

<u>Claim 1</u>

Uniformly computable in $n,\vec{\psi}$ we may decide

$$\exists m \geq n\ h(m,e,\vec{\psi}) \neq \{e\}(\vec{\psi})$$

<u>Proof</u>: Let n be given. If $h(n,e,\vec{\psi}) \neq \{e\}(\vec{\psi})$ we know that $\exists m \geq n\ h(m,e,\vec{\psi}) \neq \{e\}(\vec{\psi})$. So assume that $h(n,e,\vec{\psi}) = \{e\}(\vec{\psi})$.

Let γ be the following functional.

If $\exists m > n\ (h(m,e,\vec{\psi}) \neq \{e\}(\vec{\psi}))$ let $\gamma(\varphi) = h(m-1,e_1,\varphi,\vec{\psi})$ for the least such m

Otherwise let $\gamma(\varphi) = \{e_1\}(\varphi,\vec{\psi})$

γ will be computable in $\vec{\psi}, n$ by the following argument. If $E(e_1, \varphi, \vec{\psi}) > 0$ we may use ${}^2E$ to compute $\gamma(\varphi)$ directly.

If $E(e_1, \varphi, \vec{\psi}) = 0$, let $m_0 = M(e_1, \varphi, \vec{\psi})$

If $\exists m(n < m \leq m_0 \wedge h(m, e, \vec{\psi}) \ast h(n, e, \vec{\psi}))$ then $\gamma(\varphi) = h(m-1, e_1, \varphi, \vec{\psi})$ for the least such m. Otherwise $\gamma(\varphi) = h(m_0, e_1, \varphi, \vec{\psi}) = \{e_1\}(\varphi, \vec{\psi})$.

But then

$$\exists m > n \ h(m, e, \vec{\psi}) \ast h(n, e, \vec{\psi}) \leftrightarrow \psi_1(\gamma) \ast \{e\}(\vec{\psi}).$$

$\quad\quad\quad\quad\quad\quad\quad\quad\quad\quad\quad\quad\quad\quad\quad\quad\quad\quad\quad$ □ Claim 1

Claim 2

Uniformly computable in $\vec{\psi}$ we may decide

$$\exists n \ \forall m \geq n \ \{e\}(\vec{\psi}) = h(m, e, \vec{\psi})$$

Proof: By Claim 1 we may decide $\exists n \geq 1 \ h(e, n, \vec{\psi}) \ast \{e\}(\vec{\psi})$. So assume that there is such n. In case there is a largest one we want to iso· late it.

Let ν be the following functional

$\nu(\varphi) = \{e_1\}(\varphi, \vec{\psi})$ if for arbitrary large n

$\quad\quad h(n, e, \vec{\psi}) \ast \{e\}(\vec{\psi})$

$\nu(\varphi) = h(n-1, e_1, \varphi, \vec{\psi})$ if n is the last number such that

$\quad\quad h(n, e, \vec{\psi}) \ast \{e\}(\vec{\psi})$

Using Claim 1 and the method of Claim 1 we see that ν is computable in ψ.

But then

$$\exists n \ \forall m \geq n \ \{e\}(\vec{\psi}) = h(m, e, \vec{\psi}) \leftrightarrow \psi_1(\nu) \ast \{e\}(\vec{\psi})$$

$\quad\quad\quad\quad\quad\quad\quad\quad\quad\quad\quad\quad\quad\quad\quad\quad\quad\quad\quad$ □ Claim 2

If $\exists n \ \forall m \geq n \ \{e\}(\vec{\psi}) = h(m, e, \vec{\psi})$ we let $E(e, \vec{\psi}) = 0$ and we let

$$M(e, \vec{\psi}) = \mu n \ \forall m \geq n \ \{e\}(\vec{\psi}) = h(m, e, \vec{\psi})$$

By Claim 1 M is computable.

If $\forall n \ \exists m \geq n \ \{e\}(\vec{\psi}) \ast h(m, e, \vec{\psi})$ we will show that we may compute ${}^2E$ from $\vec{\psi}$. We let $E(e, \vec{\psi})$ be the index we may derive from this algorithm for ${}^2E$.

Let f be given. Let δ_f be the following functional

$$\delta_f(\varphi) = \begin{cases} h(m-1, e_1, \varphi, \vec{\psi}) \text{ for the least } m \text{ such that} \\ \quad\quad \exists n \leq m \ f(n) = 0 \wedge h(m, e, \vec{\psi}) \ast \{e\}(\vec{\psi}) \\ \text{if there is such } m \\ \{e_1\}(\varphi, \vec{\psi}) \text{ otherwise.} \end{cases}$$

Clearly δ_f is uniformly computable from $f,\vec{\psi}$ and

$$\exists n\, f(n) = 0 \leftrightarrow \psi_1(\delta_f) \neq \{e\}(\vec{\psi})$$

This ends the proof of the Theorem.

□

Remark 6.4

<u>a</u>   The method of this proof goes back to Grilliot [14]. Wainer [45] was first in using this method for analyzing all computations in a functional. This proof was first given in Normann [36].

<u>b</u>   Theorem 6.3 shows that ^2E is deeply connected with discontinuity. Moreover, if ^2E is not computable in ϕ then the continuity-properties of computations in ϕ are not disturbed by the fact that some subcomputations would involve computations in tremendously powerful functionals such as ^3E (see section 1.3). This is illustrated in the two main corollaries of Theorem 6.3 (6.5 and 6.8).

Corllary 6.5

Let $\vec{\psi}$ be a list of functionals such that ^2E is not computable in $\vec{\psi}$. Then there is a function $h_{\vec{\psi}}$ primitive recursive in $\vec{\psi}$ such that for any $f \in Tp(1)$

f is computable in $\vec{\psi}$ if and only if there is an r.e. ($h_{\vec{\psi}}$)-set D computable in $\vec{\psi}$ such that f is recursive in $h_{\vec{\psi}}, D$.

<u>Proof</u>: Let $h_{\vec{\psi}}(n,e,x) = h(n,e,x,\vec{\psi})$. Let f be computable in $\vec{\psi}$ and choose e such that for all $x \in \omega$

$$f(x) = \{e\}(x,\vec{\psi})$$

Since ^2E is not computable in $\vec{\psi}$ it follows from Theorem 6.3 that for all x

$$\{e\}(x,\vec{\psi}) = \lim_{n\to\infty} h(n,e,x,\vec{\psi}) = \lim_{n\to\infty} h_{\vec{\psi}}(n,e,x)$$

and that $M(e,x,\vec{\psi})\!\downarrow$ for all $x \in \omega$.

Let $D_e = \{(n,x); \exists m>n\ h_{\vec{\psi}}(m,e,x) \neq h_{\vec{\psi}}(n,e,x)\}$. Clearly D_e is r.e. ($h_{\vec{\psi}}$) and f is recursive in $D_e, h_{\vec{\psi}}$ by the following algorithm for $f(x)$:

Find the least n such that $(n,x) \notin D_e$. Then $f(x) = h_{\vec{\psi}}(n,e,x)$.

Moreover D_e is computable in $\vec{\psi}$ by

$$(n,x) \in D_e \leftrightarrow \exists m \leq M(e,x,\vec{\psi})(n<m \wedge h(m,e,x,\vec{\psi}) \neq h(n,e,x,\vec{\psi}))$$

This establishes all parts of the corollary.

□

Remark 6.6

This corollary is the analogue of Corollary 5.53 for Kleene-computations. It was first proved for type-2 functionals in Wainer [45].

As we remarked in the beginning of this section this proof works for all type-structures $<T_k>_{k\in\omega}$ closed under computations. An immediate consequence is

Corollary 6.7

If $T = <T_k>_{k\in\omega}$ is a type-structure closed under computations and $^2E \upharpoonright T_1 \in T_2$, then for all $\psi \in T$

1-sc(ψ) is generated by its r.e. elements modulo some $h \in 1$-sc(ψ).

□

These results can be used to characterize the functionals ψ in which ^2E is computable. Our next corollary was proved for $\psi \in Tp(2)$ by Grilliot [14] and for $\psi \in Tp(3)$ by Bergstra [1]. Bergstra conjectured the general result.

Corollary 6.8

Let ψ be a functional. Then the following are equivalent

i ^2E is computable in ψ

ii   1-sc(ψ) is closed under ordinary jump.

Proof:

i →ii: is trivial. ^2E "computes" quantifiers.

ii →i: If i does not hold then there is an $h \in 1$-sc(ψ) such that every $f \in 1$-sc(ψ) is recursive in the jump of h. Thus the double jump of h cannot be in 1-sc(ψ).

□

Corollary 6.9

There is a countable initial segment of the degrees that is not the 1-section of any functional.

Proof: Let $0^{(n)}$ be the n'th jump of the degree of the recursive functions. Let

$$f \in A \leftrightarrow \exists n(f \text{ is recursive in } 0^{(n)})$$

Assume that $A = 1\text{-sc}(\psi)$. ^2E cannot be computable in ψ since then

$$A \subsetneq 1\text{-sc}(^2E) \subseteq 1\text{-sc}(\psi)$$

On the other hand A is closed under jump. So ^2E is computable in ψ by Corollary 6.8.

This is impossible so A is not the 1-section of any functional.

□

Remark 6.10

Using some advanced methods from the theory of Turing degrees one may use Corollary 6.5 to show that there is an initial segment Λ of the Δ_2^0-degrees that is not the 1-section of any functional. If A is a topless, wellfounded initial segment of degrees then $A = 1\text{-sc}(\psi)$ will contradict the density theorem.

Corollary 6.11

Let $\psi \in Ct(k)$, $k \geq 3$. Then there is an $h \in 1\text{-sc}(\psi)$ such that

<u>i</u> $1\text{-sc}(\psi)$ is generated by its r.e. degrees modulo h

<u>ii</u> $1\text{-sc}(\psi)$ is $\pi_{k-2}^1(h)$

Proof:

<u>i</u> is just a special case of Corollary 6.7. By Theorem 5.47
$$2\text{-en}(\psi) = \pi_{k-2}^1(h). \quad \text{So } C = \{<e,n,m>; \{e\}(n,\psi) \simeq m\} \text{ is } \pi_{k-2}^1(h).$$
Then
$$1\text{-sc}(\psi) = \{f : \exists e \ \forall n <e,n,f(n)> \in C\}$$
will be $\pi_{k-2}^1(h)$.
(Note that the h here and the trace of ψ used in 5.47 are arithmetical in each other.)

□

Remark 6.12

In section 6.3 we will show that the necessary conditions for A to be a 1-section of a countable functional given in Corollary 6.11 are indeed sufficient.

6.2 The 1-section of a type-2 functional

Recall that a functional F of type-2 is called <u>normal</u> if ^2E is computable in F. In particular no continuous functionals will be normal. 1-sections of normal functionals of type-2 are well treated elsewhere (Sacks [40], Fenstad [11]) and will not be dealt with here.

When it comes to computations in a non-normal functional F of type-2 we know from Theorem 6.3 that F 'behaves' like a continuous functional. If we take a closer look at the proof of 6.3 we see how we get from $\lambda x\{e\}(x,F)$ to $F(\lambda x\{e\}(x,F))$.

We compute a modulus-function for the sequence

$$<h(n,e,x,F)>_{n\in\omega}$$

and find a modulus for

$$<h(n,e_1,F)>_{n\in\omega}$$

where $\{e_1\}(F) = F(\lambda x\{e\}(x,F))$.

If we had another operator computing the modulus for $<h(n,e_1,F)>_{n\in\omega}$ from the modulus function for $<h(n,e,x,F)>_{n\in\omega}$ we could use this and h to recover $\{e_1\}(F)$.

What we actually want to do is to construct an operator sending the r.e. (h)-set

$$\{(x,n)\ \exists m>n\ h(n,e,x,F) \neq h(m,e,x,F)\}$$

to the r.e. (h)-set

$$\{n\ ;\ \exists m >n\ h(n,e_1,F) \neq h(m,e_1,F)\}.$$

In section 4.2 we saw how to construct such local 'jumps' from one r.e. set to another, the $F_e^{e'}$-functionals.

We will now define an operator J_F which actually is an effective family of such $F_e^{e'}$.

Definition 6.13

Let $F \in Tp(2)$.

<u>a</u>   Let $h_F(n,e,<\vec{a}>) = h(n,e,\vec{a},F)$ where $\vec{a}$ is a finite sequence of natural numbers.

<u>b</u>   Let

$$Mod(e,\vec{a},\sigma) \leftrightarrow \forall x,j < lh(\sigma)[j>\sigma(x) \rightarrow$$
$$h_F(j,e,<x,\vec{a}>) = h_F(\sigma(x),e,<x,\vec{a}>)]$$

<u>c</u>   For any index e let e' be primitive recursively computable
from e such that

$$\{e'\}(\vec{a},F) \simeq F(\lambda x\{e\}(x,\vec{a},F))$$

<u>d</u>   Let

$$J_F(\alpha,<e,<\vec{a}>,n>) = \begin{cases} 0 & \text{if } \exists m > n\,(h_F(m,e',\vec{a}) \neq h_F(n,e',\vec{a}) \\ & \qquad\qquad \wedge \text{Mod}(e,\vec{a},\bar{\alpha}(m))) \\ \\ 1 & \text{otherwise} \end{cases}$$

Remark 6.14

<u>a</u> $\text{Mod}(e,\vec{a},\sigma)$ means that σ looks like the beginning of a
modulus-function for the sequence
$$k_i(x) = h_F\,(i,e,\,<x,\vec{a}>)$$
when only numbers $< \text{lh}(\sigma)$ are checked (cfr. the definition of $F_e^{e'}$)

<u>b</u>   If α is a modulus for $\{k_i\}_{i\in\omega}$ as described in <u>a</u> then
$J_F(\alpha,<e,<\vec{a}>,n>)$ will tell us whether we finally meet the value
$$F(\lambda x\{e\}(x,\vec{a},F))$$

in $h_F(n,e',\vec{a})$, i.e. we may then use J_F to compute a modulus
for the approximation of the computation
$$F(\lambda x\{e\}(x,\vec{a},F))$$

Lemma 6.15

   J_F is continuous and the Kalmar Rank of J_F is $\leq \omega$.

<u>Proof</u>: J_F is clearly computable in the jump of h_F , which is a
type-1-object, so J_F is continuous.

   To show that the Kalmar Rank of J_F is $\leq \omega$ it is sufficient to
show that for each $<e,<\vec{a}>,n>$ the Kalmar Rank of $J_{F,<e,<\vec{a}>,n>}(\alpha) =$
$J_F(\alpha,<e,\vec{a}>,n>)$ is finite.

   Now if $\exists m > n\, h_F(m,e,<\vec{a}>) \neq h_F(n,e,<\vec{a}>)$ choose the minimal such m .
If $\text{Mod}(e,\vec{a},\bar{\alpha}(m))$ we will have $J_{F,<e,<\vec{a}>,n>}(\alpha) = 0$, otherwise 1 .
So the value is decided by $\bar{\alpha}(m)$ and the Kalmar Rank is finite.

   If $\forall m > n\, h_F(m,e,<\vec{a}>) = h_F(n,e,<\vec{a}>)$ then $J_{F,<e,\vec{a}>,n>}$ is constant
1 and the Kalmar Rank is finite.

□

Lemma 6.16

There is a partial function M_F computable in J_F such that if $\{e\}(\vec{a},F)\!\downarrow$ then $M_F(e,\vec{a})\!\downarrow$ and $\forall m \geq M_F(e,\vec{a})(h_F(m,e,\vec{a}) = \{e\}(\vec{a},F))$.

Proof: We show that M from Theorem 6.3 when restricted to computations in F and natural numbers, is computable in J_F.

M_F is defined using the recursion theorem and the construction is given by induction on the length of the computation $\{e\}(\vec{a},F)$. The only nontrivial case is case 8 :

$$\{e\}(\vec{a},F) = F(\lambda x\{e_1\}(x,\vec{a},F))$$

By the induction hypothesis we assume that $\alpha = \lambda x\, M_F(e_1,x,\vec{a})$ is a modulus function for the sequence

$$\{\lambda x\, h_F(n,e_1,<x,\vec{a}>)\}_{n\in\omega}$$

Then for all t $\mathrm{Mod}(e_1,\vec{a},\bar{\alpha}(t)\,)$.

Let $M_F(e,\vec{a}) = \mu n(J_F(\alpha,<e_1,\vec{a},n>) = 1)$. By the definition of J_F it follows that $M_F(e,\vec{a})$ has the required property.

$\square$

Lemma 6.17

Let $F \in \mathrm{Tp}(2)$, $f \in \mathrm{Tp}(1)$. If ^2E is not computable in F, f then $J_F \restriction 1\text{-sc}(F,f)$ is computable in F.

Proof: Assume that $\alpha \in 1\text{-sc}(F,f)$ and let $<e,\vec{a},n>$ be given. We want to give instructions how to compute $J_F(\alpha,<e,\vec{a},n>)$. The proof follows the same pattern as the proof of Theorem 6.3.
Recall that

$$J_F(\alpha,<e,\vec{a},n>) = \begin{cases} 0 & \text{if } \exists m > n\ (h_F(m,e',\vec{a}) \neq h_F(n,e',\vec{a}) \\ & \qquad\qquad \wedge \mathrm{Mod}(e,\vec{a},\bar{\alpha}(m))) \\ 1 & \text{otherwise} \end{cases}$$

Let $g_n(x)$ be defined by:

If there is an m such that $n < m \leq \max\{\alpha(x),x+1\}$,

$$h_F(m,e',\vec{a}) \neq h_F(n,e',\vec{a}) \quad \text{and} \quad \mathrm{Mod}(e,\vec{a},\bar{\alpha}(m))$$

then let $g_n(x) = h_F(m_0-1,e,x,\vec{a})$ where m_0 is the least such m.

If there is no such m let $g_n(x) = h_F(\alpha(x),e,x,\vec{a})$. g_n is clearly recursive in α uniformly in $n,e,\vec{a}$.

Claim 1

If there is an $m > n$ such that

$$h_F(m,e',\vec{a}) \neq h_F(n,e',\vec{a}) \wedge Mod(e,\vec{a},\bar{a}(m))$$

then

$$g_n(x) = h_F(m_0-1,e,x,\vec{a})$$

where m_0 is the least such m.

Proof: For each x we look at two cases:

__i__ $m_0 \leq \max\{\alpha(x),x\}$. Then
$g_n(x) = h_F(m_0-1,e,x,\vec{a})$ directly by the instruction.

__ii__ $m_0 > \max\{\alpha(x),x\}$. Then $g_n(x) = h_F(\alpha(x),e,x,\vec{a})$. But since
$Mod(e,\vec{a},\bar{a}(m_0))$ and $x < lh(\bar{a}(m_0)) = m_0$ we see that

$$\forall j < m_0 (j \geq \alpha(x) \rightarrow h_F(j,e,x,\vec{a}) = h_F(\alpha(x),e,x,\vec{a}))$$

If we let $j = m_0-1$ we see that the claim holds.

▢ Claim 1

Claim 2

If for all $m > n$ we have

$$h_F(m,e',\vec{a}) = h_F(n,e',a) \vee \neg Mod(e,\vec{a},\bar{a}(m))$$

then $g_n(x) = h(\alpha(x),e,x,\vec{a})$ for all x.

Proof: Immediate from the definition of $g_n(x)$.

▢ Claim 2

If we can decide computably in F which of the premisses of
Claims 1 and 2 is true we can give the desired algorithm for
$J_F(\alpha,<e,\vec{a},n>)$.

We regard two cases.

Claim 3

If $h_F(n,e',\vec{a}) = F(\lambda x\, h_F(\alpha(x),e,x,\vec{a}))$
then

$$J_F(\alpha,<e,\vec{a},n>) = \begin{cases} 0 & \text{if } h_F(n,e',\vec{a}) \neq F(g_n) \\ 1 & \text{if } h_F(n,e',\vec{a}) = F(g_n) \end{cases}$$

Proof: If $J_F(\alpha,<e,\vec{a},n>) = 0$ then m_0 exists so by Claim 1 $g_n(x) = h_F(m_0-1,e,x,\vec{a})$ and by definition of m_0 we see that

$$h_F(n,e',\vec{a}) \neq h_F(m_0,e',\vec{a}) = F(\lambda x\ h_F(m_0-1,e,x,\vec{a})) = F(g_n)$$

If $J_F(\alpha,<e,\vec{a},n>) = 1$ then m_0 does not exist and by Claim 2 $g_n(x) = h_F(\alpha(x),e,x,\vec{a})$. So

$$F(g_n) = F(\lambda x\ h_F(\alpha(x),e,x,\vec{a})) = h_F(n,e',\vec{a})$$

□ Claim 3

Now suppose that $h_F(n,e',\vec{a}) \neq F(\lambda x\ h_F(\alpha(x),e,x,\vec{a}))$, Let

$$\beta_m(x) = \begin{cases} h_F(\alpha(x),e,x,\vec{a}) & \text{if } x,\alpha(x) < m \\ h_F(m-1,e,x,\vec{a}) & \text{otherwise} \end{cases}$$

Then $\lim_{m\to\infty} \beta_m = \lambda x\ h_F(\alpha(x),e,x,\vec{a})$ and the sequence $<\beta_m>$ with limit is computable in α,F . Since α is computable in f and ^2E is no computable in f,F we see that

$$F(\lambda x\ h_F(\alpha(x),e,x,\vec{a})) = \lim_{m\to\infty} F(\beta_m)$$

Let $m_1 = \mu m > n\ (h_F(n,e',\vec{a}) \neq F(\beta_m))$.

Claim 4

$$J_F(\alpha,<e,\vec{a},n>) = 0 \leftrightarrow \text{Mod}(e,\vec{a},\bar{\alpha}(m_1)) .$$

Proof:
→: Assume $J_F(\alpha,<e,\vec{a},n>) = 0$.
Let $m_0 = \mu m(h_F(n,e',\vec{a}) \neq h_F(m,e',\vec{a}) \wedge \text{Mod}(e,\vec{a},\bar{\alpha}(m)))$.
Since $\text{Mod}(e,\vec{a},\bar{\alpha}(m_0))$ holds we have for all $j \le m_0$

$$\beta_j = \lambda x\ h_F(j-1,e,x,\vec{a})$$

Hence $F(\beta_j) = h_F(j,e',\vec{a})$ for $j \le m_0$. Thus $m_1 \ge m_0$. But $m_1 \le m_0$ since

$$h_F(n,e',\vec{a}) \neq h_F(m_0,e',\vec{a}) = F(\beta_{m_0})$$

So $m_1 = m_0$. But then $\text{Mod}(e,\vec{a},\bar{\alpha}(m_1))$ holds.

←: Assume that $\text{Mod}(e,\vec{a},\bar{\alpha}(m_1))$ holds.
By choice of m_1 we see that if $n < j < m_1$ then

$$F(\beta_j) = h_F(n,e',\vec{a}) \quad \text{while} \quad F(\beta_{m_1}) \neq h_F(n,e',\vec{a})$$

Since $\text{Mod}(e,\vec{a},\bar{\alpha}(m_1))$ holds we have for all $j \le m_1$

$$F(\beta_j) = h_F(j,e',\vec{a})$$

Thus m_1 is minimal such that

$$h_F(m_1,e',\vec{a}) \neq h_F(n,e',\vec{a}) \wedge \text{Mod}(e,\vec{a},\bar{a}(m_1))$$

But by definition of J_F then

$$J_F(\alpha,<e,\vec{a},n>) = 0$$

□ Claim 4

Using these claims we obtain an algorithm for computing $J_F(\alpha,<e,\vec{a},n>)$. If $h_F(n,e',\vec{a}) = F(\lambda x\ h_F(\alpha(x),e,x,\vec{a}))$ then we use Claim 3 to compute $J_F(\alpha,<e,\vec{a},n>)$. If $h_F(n,e',\vec{a}) \neq F(\lambda x\ h_F(\alpha(x),e,x,\vec{a}))$ we construct the β_m's and find m_1 and then we may use Claim 4 to compute $J_F(\alpha,<e,\vec{a},n>)$.

□

The previous lemmas and the following theorem are all due to S. Wainer. (Normann-Wainer [35].)

Theorem 6.18

Assume that ^2E is not computable in F and let J_F be constructed as in Definition 6.13 $\underline{d}$. Then

$$1\text{-sc}(F) = 1\text{-sc}(J_F,h_F)$$

Proof:

$\subseteq$: If $f = \lambda x\ \{e\}(x,F)$ then $f(x)$ may be computed from J_F,h_F by the following algorithm: Let M_F be partial recursive in J_F as in Lemma 6.16. Let $m = M(e,x)$. Then $f(x) = h_F(m,e,\vec{a})$.

$\supseteq$: Since h_F is primitive recursive in F it is sufficient to show that $J_F \upharpoonright 1\text{-sc}(F)$ is computable in F (see Theorem 4.11). But by Lemma 6.17 this is the case.

□

Corollary 6.19

If F is a type-2 functional such that ^2E is not computable in F, then there is a continuous G of type-2 with Kalmar Rank $\leq \omega$ such that $1\text{-sc}(F) = 1\text{-sc}(G)$

Proof: Let $G = <J_F,h_F>$. By Lemma 6.15 $G \in \text{Ct}(2)$ and $\text{KR}(G) \leq \omega$.

□

Corollary 6.20

If $F \in \text{Ct}(2)$ then there is a $G \in \text{Ct}(2)$ such that G is computable in F, $\text{KR}(G) \leq \omega$ and $1\text{-sc}(F) = 1\text{-sc}(G)$.

Proof: Let $G = \langle J_F, h_F \rangle$. Since F is continuous it follows from Lemma 6.17 that G is computable in F . The other properties are proved as before.

□

If we are mainly interested in 1-sections and 2-envelopes, Theorem 6.18 with corollaries gives us some sort of normal form for a continuous functional of type-2. In particular we know that all 1-sections are obtained from functionals with Kalmar Rank $\leq \omega$ i.e. a class of functionals being covered by a Δ^0_ω-class of associates.

Moreover if we want to construct 1-sections with some special property we know that if we will succeed at all we will probably succeed by using functionals $F_e^{e'}$ as building blocks. This explains why the $F_e^{e'}$-method has been so fruitful (Bergstra [1], Bergstra-Wainer [4], Normann [32] and Normann-Wainer [35]).

In the rest of this section we will give an illustration of how this method works. For a further discussion see Normann-Wainer [35].

Lemma 6.21

There is an r.e. set $B \subseteq \mathbb{N}^2$ such that for each n , B_n is not recursive in B_{-n} where

$$B_n = \{\langle n,a \rangle ; \langle n,a \rangle \in B\}$$

$$B_{-n} = \{\langle m,a \rangle \in B; m \neq n\}$$

Proof: The proof of this lemma is by an injury argument and not within the scope of this book. The method is essentially the same as the one used to solve Post's problem, you just satisfy suitable requirements for each n . For further details on priority arguments we leave the reader to any textbook in ordinary degree-theory e.g. Sacks [39], Shoenfield [43] or Rogers [38].

□

From now on and throughout this section we let B be fixed as in Lemma 6.21 , $\prec$ a recursive partial ordering of a recursive subset D of the natural numbers. When $n \in D$ we let

$$Y_n = \{\langle a,m \rangle ; \langle a,m \rangle \in B \wedge m \in D \wedge m \prec n\}$$

$$X_n = \{\langle a,m \rangle ; \langle a,m \rangle \in B \wedge m \in D \wedge m \preceq n\}$$

Both Y_n and X_n are r.e. sets. Y_n is recursive in X_n uniformly in n while X_n is not recursive in Y_n .
Without loss of generality we may assume that the set

$$\{n \in D \; ; \; n \text{ is } \prec\text{-minimal}\}$$

is recursive.

Let p and q be primitive recursive functions such that $Y_n = W_{p(n)}$, $X_n = W_{q(n)}$ for all $n \in D$.

We recall the following definitions from section 4.2 :

$$Mod(e,\sigma) \leftrightarrow \forall x < lh(\sigma)(\exists t < lh(\sigma) \ T(e,x,t) \rightarrow \exists t \leq \sigma(x)T(e,x,t))$$

$$F_e^{e'}(x,\alpha) = \begin{cases} 0 & \text{if } \exists t \ T(e',x,t) \wedge Mod(e,\bar{\alpha}(t)) \\ 1 & \text{otherwise} \end{cases}$$

$G_e^{e'}$ is computed by the following partial algorithm:

Find the least s such that $\neg Mod(e,\bar{\alpha}(s))$. Then

$$G_e^{e'}(x,\alpha) = \begin{cases} 0 & \text{if } \exists t < s \ T(e',x,t) \\ 1 & \text{if } \forall t < s \neg T(e',x,t) \end{cases}$$

In section 4.2 we showed that $G_e^{e'}$ is a computable subfunction of $F_e^{e'}$ and $G_e^{e'}(x,\alpha)$ is defined whenever W_e is not recursive in α. Now let F be defined by

$$F(n,x,\alpha) = \begin{cases} F_{p(n)}^{q(n)}(x,\alpha) & \text{if } n \in D \\ 0 & \text{if } n \notin D \end{cases}$$

Lemma 6.22

Let F be defined as above. If $n \in D$ is in the $\prec$-well-founded part of D then X_n is uniformly computable in F,n.

Proof: Formally we will use the recursion theorem to give a uniform index for the computation of X_n from F and n . Our argument is by induction on the well-founded part of n .
We will regard two cases.

$\underline{1}$ n is minimal. Then Y_n is empty and $\alpha(x) = 0$ for all x will be a modulus-function for Y_n . But from Lemma 4.16 $\underline{d}$ we know how to compute X_n from a modulus function for Y_n by the help of $F_{p(n)}^{q(n)}$.

$\underline{2}$ n is not minimal. By an induction-hypothesis X_m is uniformly computable in m,F for $m \prec n$. But $Y_n = \bigcup_{m \prec n} X_m$ so Y_n is computable in F .
We then use F to compute X_n . By the recursion theorem we may assume that we use the same index.

□

Now for each $n \in D$ define

$$H_n(m,x,\alpha) = \begin{cases} F(m,x,\alpha) & \text{if } m \in D \text{ and } \neg n \leq m \\ G_{p(m)}^{q(m)}(x,\alpha) & \text{if } n \leq m \\ 0 & \text{if } m \notin D \end{cases}$$

H_n is a partial subfunction of F.

Lemma 6.23

<u>a</u>   For each $n \in D$ H_n is computable in B_{-n}.

<u>b</u>   If Y_n is not computable in α then $H_n(m,x,\alpha) = F(m,x,\alpha)$.

Proof:

<u>a</u>   Assume that $m \in D$ and that x and α are arbitrary. We will show how to compute $H_n(m,x,\alpha)$.

If $n \leq m$ then we apply the recursive $G_{p(n)}^{q(m)}$ so assume $\neg(n \leq m)$. Then X_m is recursive in B_{-n} uniformly in m. Now $F_{p(m)}^{q(m)}$ is uniformly computable in X_m,m and $H_n(m,x,\alpha) = F_{p(m)}^{q(m)}(x,\alpha)$. This shows how to compute $H_n(m,x,\alpha)$ from B_{-n}.

<u>b</u>   Assume that Y_n is not recursive in α. If $m \notin D$ or $\neg(n \leq m)$ then $H_n(m,x,\alpha) = F(m,x,\alpha)$ by definition. If $n \leq m$ we cannot have that Y_m is recursive in α since Y_n is recursive in Y_m. But then

$$F(m,x,\alpha) = F_{p(m)}^{q(m)}(x,\alpha) = G_{p(m)}^{q(m)}(x,\alpha) = H_n(m,x,\alpha)$$

□

Lemma 6.24

Let $n \in D$ and assume that n is not in the well-founded part of $\leq$. Let $\vec{a}$ be a finite list of natural numbers and assume that $\{e\}(F,\vec{a}) \simeq k$. Then $\{e\}(H_n,\vec{a}) \simeq k$.

Proof: We prove this by induction on the length of the computation $\{e\}(F,\vec{a}) \simeq k$.

The only non-trivial case is when we apply F, i.e.

$$\{e\}(F,\vec{a}) = F(m,x,\lambda y\{e_1\}(F,y,\vec{a}))$$

Now choose $n_1 < n$ such that n_1 is not in the well-founded part. By the induction hypothesis we have that for all y

$$\{e_1\}(F,y,\vec{a}) \simeq \{e_1\}(H_{n_1},y,\vec{a})$$

Let $\alpha(y)$ be this common value. Then α is computable in H_{n_1} so by Lemma 6.23 $\underline{a}$ α is recursive in B_{-n_1}. But then B_{n_1} is not recursive in α, and since B_{n_1} is recursive in Y_n we cannot have that Y_n is recursive in α. By Lemma 6.23 $\underline{b}$

$$H_n(m,x,\alpha) = F(m,x,\alpha)$$

But $\alpha(y)$ is also $\{e_1\}(H_n,y,\vec{a})$ so

$$\{e\}(F,\vec{a}) = F(m,x,\alpha) = H_n(m,x,\alpha) = \{e\}(H_n,\vec{a})$$

□

As a consequence of Lemma 6.22 and Lemma 6.24 we see that X_n is computable in F if and only if n is in the well-founded part of D.

One of the more general consequences of this method is given in the following result:

Theorem 6.25

Let $A \subseteq \omega$ be a π^1_1-set. Then there is a primitive recursive function ρ and an $F \in Ct(2)$ computable in $0'$ such that

$$\forall n \in \omega \ (W_{\rho(n)} \in 1\text{-sc}(F) \leftrightarrow n \in A)$$

Proof: Let R be recursive such that

$$n \in A \leftrightarrow \forall f \ \exists t \ R(n,\bar{f}(t))$$

Let $D = \{\langle n,\sigma \rangle ; \neg R(n,\sigma)\}$ with $\langle n,\sigma \rangle \prec \langle m,\tau \rangle$ if $n = m$ and σ is an extension of τ.

Let $\rho(n)$ be an index for $X_{\langle n,\langle\rangle\rangle}$. Clearly

$$n \in A \leftrightarrow \langle n,\langle\rangle\rangle \text{ is in the well-founded part of } D.$$

The theorem then follows.

□

The following corollary was first proved in Normann [32] answering the following problem raised by Stan Wainer. In Wainer [46] hierarchies are defined for recursion in arbitrary functionals of type-2. Wainer asked if there are continuous functionals for which his hierarchy produces new recursive functions arbitrarily high. Corollary 6.26 provides an example.

Corollary 6.26

There is a functional F computable in $0'$ such that

$$1\text{-sc}(F) \in \pi^1_1 \smallsetminus \Sigma^1_1$$

Proof: Choose $A \in \pi_1^1 \smallsetminus \Sigma_1^1$ in Theorem 6.25.
Now

$$n \in A \leftrightarrow W_{\rho(n)} \in 1\text{-sc}(F)$$

so $1\text{-sc}(F)$ cannot be Σ_1^1. But we know that $1\text{-sc}(F)$ is always π_1^1.

□

There are some other applications of Theorem 6.25 given in Normann-Wainer [35].

In the next section we will give an index-free characterization of 1-sections of countable functionals of type > 2. In Wainer [45] there is a characterization of 1-sections of type-2 functionals using indices. Is there an index-free characterization, an analogue to Sacks [40] ? A simpler problem could be:

Is there a π_1^1-segment A of the degrees recursively generated by its r.e. members such that A is not a 1-section ?

6.3 The 1-section of a higher type functional

There are several problems left open concerning the nature of the 1-sections of type-2 functionals. We know that they are generated by their r.e. degrees and we know how to construct some of them. But there are still problems and the solutions of these problems will add new insight in the mechanism of Kleene-computations.

Moving up in type the situation is much clearer. We will give a characterization of all 1-sections of continuous functionals of type > 2 avoiding all mentioning of hierarchies. This section is mostly devoted to the proof of the following theorem:

Theorem 6.27
    Let $A \subseteq Tp(1)$, $k \geq 1$. The following are then equivalent:

<u>a</u>   There is a functional $\Psi \in Ct(k+2)$ such that

$$A = 1\text{-sc}(\Psi)$$

<u>b</u>   There is a function $h \in A$ such that

<u>i</u> $A \in \pi_k^1(h)$

<u>ii</u>   If $f_1, \ldots, f_n \in A$ and g is recursive in $<f_1, \ldots, f_n>$ then $g \in A$.

<u>iii</u>   If $f \in A$ then there is an r.e.(h)-set B with characteristic function in A such that f is recursive in B, h .

Proof: $\underline{a} \rightarrow \underline{b}$ is a consequence of Corollary 5.46 and Corollary 6.7.

When we prove $\underline{b} \rightarrow \underline{a}$ we drop h in the argument. The argument we give can be relativized to h without changes. So we assume that A is π_k^1, closed under recursion in finite lists and recursively generated by its r.e. elements. We want to construct $\Psi \in Ct(k+2)$ such that $A = 1\text{-}sc(\Psi)$.

We will use Theorem 5.22 $\underline{a}$, or rather Corollary 5.23 $\underline{a}$. So we fix a primitive recursive set $\{\varphi_n\}_{n\in\omega}$ dense in $Ct(k)$. For $\psi \in Ct(k+1)$ let $h_\psi(n) = \psi(\varphi_n)$ be the trace of ψ.
Let $B \subseteq \omega$ be π_k^1. Then there is a primitive recursive relation R such that

$$m \in B \leftrightarrow \exists \psi \in Ct(k+1) \, \forall n \, R(m, \bar{h}_\psi(n), n)$$

where we may choose ψ μ-recursive uniformly in m if $m \in B$.

For this for the moment rather arbitrary choice of B and R we let

$$\phi(e,k,\psi) = \begin{cases} 1 & \text{if } \exists s (T(e,k,s) \wedge \forall n \leq s \, R(e, \bar{h}_\psi(n), n)) \\ 0 & \text{otherwise} \end{cases}$$

Claim 1

ϕ is computable in $0'$ and if $e \in B$ then W_e is μ-recursive in ϕ.

Proof: From $0'$ we may decide if $\exists s \, T(e,k,s)$. If there is no such s, then $\phi(e,k,\psi) = 0$. Otherwise let s be minimal such that $T(e,k,s)$. Then

$$\phi(e,k,\psi) = \begin{cases} 1 & \text{if } \forall n \leq s \, R(e, \bar{h}_\psi(n), n) \\ 0 & \text{if } \exists n \leq s \, \neg R(e, \bar{h}_\psi(n), n) \end{cases}$$

If $e \in B$ we let ψ be μ-recursive such that
$$\forall n \, R(e, \bar{h}_\psi(n), n)$$
Then

$$k \in W_e \leftrightarrow \exists s \, T(e,k,s) \leftrightarrow \phi(e,k,\psi) = 1$$

So W_e is μ-recursive in ϕ.
$\qquad\qquad\qquad\qquad\qquad$ □ Claim 1

It follows from Claim 1 that if we choose B such that B contains an index for each r.e. element in A, and if ϕ is constructed as above then $A \subset 1\text{-}sc(\phi)$.

We also see that when $e \notin B$ then $\lambda k, \psi \phi(e,k,\psi)$ is computable since we may always find a least n such that $\neg R(e, \bar{h}_\psi(n), n)$ and in order to compute $\phi(e,k,\psi)$ we may check if $T(e,k,s)$ holds for $s < n$.

We will first prove the theorem for $k = 1$ and then for $k > 1$. The main idea is the same but there are some local differences. So assume $k = 1$.

Let $B_0 = \{e; W_e \in A\}$. Then B_0 is Π_1^1. Let C be Δ_1^1 such that all infinite arithmetical sets intersect both C and the complement of C. We leave the construction of C to the reader. Let $B = B_0 \cap C$, R and ϕ as above.

Claim 2

$A \subseteq 1\text{-sc}(\phi)$.

<u>Proof</u>: Let $W_e \in A$. Then $\{e'; W_{e'} = W_e\}$ is an infinite arithmetical subset of B_0. By choice of C there is an $e' \in C$ such that $W_{e'} = W_e$, so $e' \in B$. Then $W_e = W_{e'} \in 1\text{-sc}(\phi)$.

A is recursively generated from its r.e. elements so $A \subseteq 1\text{-sc}(\phi)$.

□ Claim 2

Now by choice of C all arithmetical subsets of B are finite. We will see that this is sufficient to give us that $1\text{-sc}(\phi) \subseteq A$.

The idea is as follows. We regard $\alpha = \lambda x \{d\}(x,\phi)$. We know that when $e \notin B$ we may replace $\phi(e,k,f)$ by an algorithm. If $e \in B$ this algorithm will give us a partial functional. We will show that with some modifications we can replace $\phi(e,k,f)$ by this algorithm for an arithmetical set of e's. The rest will be an arithmetical subset of B and thus finite. Then α will be recursive in a finite list from A and thus in A itself.

We now move over to the details. Let d be fixed and assume that $\alpha = \lambda x \{d\}(x,\phi)$ is total. In any subcomputation $\{d_1\}(\vec{\psi})$ of a computation $\{d\}(x,\phi)$, ϕ will be the only object in $\vec{\psi}$ of type > 1, the rest will be natural numbers and functions.

We say that a computation $\{d_1\}(\vec{a}, \vec{f}, \phi)$ is <u>essential</u> if all function-arguments in $\vec{f}$ are from the list $\{\varphi_n\}_{n \in \omega}$.

Claim 3

The family of essential subcomputations of $\{d\}(x,\phi)$, including $\{d\}(x,\phi)$ itself, is arithmetical with an arithmetical enumeration.

Proof: ϕ is computable in $0'$ so let β be an associate for ϕ recursive in $0'$.

If $\{d_1\}(\vec{a},\vec{f},\phi)$ is any computation we may find the value effectively in β. This means that the set of immediate subcomputations of $\{d_1\}(\vec{a},\vec{f},\phi)$ is uniformly recursive in $\beta,d_1,\vec{a},\vec{f}$.

Now $\{d_1\}(\vec{a},\vec{f},\phi)$ is an essential subcomputation of $\{d\}(x,\phi)$ if $\vec{f}$ is from $\{\varphi_n\}_{n\in\omega}$ and there is a finite list $\sigma_1,\ldots,\sigma_t$ of computation-tuples such that $\sigma_1 = \langle d,x,\phi\rangle$, $\sigma_t = \langle d_1,\vec{a},\vec{f},\phi\rangle$ and each σ_{i+1} is an immediate subcomputation of σ_i ($i<t$). Such computation-tuples may be coded by natural numbers since all function-arguments are from the enumerated list $\{\varphi_n\}_{n\in\omega}$. So we may give an arithmetical definition of the essential subcomputations of $\{d\}(x,\phi)$ when x varies over ω.

$\quad\quad\quad\quad$ □ Claim 3

We say that ϕ is used <u>non-effectively</u> at e if there is an x and an essential subcomputation

$$\{d_1\}(\vec{a},\vec{f},\phi) \simeq \phi(e,k,\lambda g\{d_2\}(\vec{a},g,\vec{f},\phi))$$

of $\{d\}(x,\phi)$ such that

$$\forall n\, R(e,\bar{h}_\psi(n),n)$$

where

$$\psi = \lambda g\,\{d_2\}(\vec{a},g,\vec{f},\phi)$$

Clearly, if ϕ is used non-effectively at e then $e\in B$. Moreover the set of e such that ϕ is used non-effectively at e is arithmetical, so this set is a finite subset D of B.

Claim 4

Let $D = \{e_1,\ldots,e_s\}$. Then α is recursive in $W_{e_1},\ldots,W_{e_s}$.

Proof: By the recursion theorem we will construct a partial functional ρ computable in $W_{e_1},\ldots,W_{e_s}$ such that if $\{d_1\}(\vec{a},\vec{f},\phi)$ is an essential subcomputation of $\{d\}(x,\phi)$ for some x then $\rho(d_1,\vec{a},\vec{f}) = \{d_1\}(\vec{a},\vec{f},\phi)$.

We define ρ by cases and as usual we describe it in practice by induction on the computation $\{d_1\}(\vec{a},\vec{f},\phi)$.

The only non-trivial case is when S8 is used,

$$\{d_1\}(\vec{a},\vec{f},\phi) \simeq \phi(e,k,\lambda g\{d_2\}(\vec{a},g,\vec{f},\phi))$$

We have two subcases:

Case 1 : $e \in D$.

If $k \notin W_e$ we let $\rho(d_1, \vec{a}, \vec{f}) = 1$ which will be the value of $\Phi(e, k, \psi)$ for any ψ .

If $k \in W_e$ we find the least s such that $T(e, k, s)$.

If

$$\forall n \leq s \, R(e, <\rho(d_2, \vec{a}, \varphi_0, \vec{f}), \ldots, \rho(d_2, \vec{a}, \varphi_{n-1}, \vec{f})>, n)$$

then let $\rho(d_1, \vec{a}, \vec{f}) = 0$, otherwise let $\rho(d_1, \vec{a}, \vec{f}) = 1$. By the induction-hypothesis $\rho(d_2, \vec{a}, \varphi_i, \vec{f})$ gives the right value so $\rho(d_1, \vec{a}, \vec{f})$ gives the right value.

Case 2 : $e \notin D$.

Then Φ is not used non-effectively at e which means that

$$\exists n \, \neg R(e, \bar{h}_\psi(n), n)$$

where $\psi = \lambda g \{d_2\}(\vec{a}, g, \vec{f}, \Phi)$.

By the induction hypothesis ρ works fine on $(d_2, \vec{a}, \varphi_i, \vec{f})$ so we use the following algorithm.

Find the least n such that

$$\neg R(e, <\rho(d_2, \vec{a}, \varphi_0, \vec{f}), \ldots, \rho(d_2, \vec{a}, \varphi_{n-1}, \vec{f})>, n)$$

Then if $\exists s < n \, T(e, k, s)$ we let $\rho(d_1, \vec{a}, \vec{f}) = 0$. Otherwise we let $\rho(d_1, \vec{a}, \vec{f}) = 1$.

It is easily seen that ρ still works.

Now $\alpha = \lambda x \, \rho(d, x)$ so α is recursive in $W_{e_1}, \ldots, W_{e_s}$.

$\square$ Claim 4

From Claim 4 it follows that if α is computable in Φ then α is recursive in a finite set from A , so $\alpha \in A$. This ends the proof when $k = 1$.

$\square$ $k = 1$

So now assume that $k > 1$.

Let C be Δ_k^1 such that all infinite Σ_{k-1}^1-subsets of ω will intersect both C and the complement of C .

Choose B_0, B and Φ as in the case for $k = 1$.

As before we easily get $A \subseteq \mu$-1-sc(Φ) and the problem is to show that 1-sc$(\Phi) \subseteq A$.

Let α be computable in Φ and choose d such that $\alpha(x) = \{d\}(x, \Phi)$ for all x .

A subcomputation $\{d_2\}(\vec{\psi}, \Phi)$ of $\{d\}(x, \Phi)$ is called essential if all functionals in $\vec{\psi}$ of type-k are from the list $\{\varphi_n\}_{n \in \omega}$. We pick canonical recursive associates $\{\gamma_n\}_{n\epsilon\omega}$ for $\{\varphi_n\}_{n\epsilon\omega}$.

Claim 5

The set of essential subcomputations of $\{d\}(x,\phi)$ is Σ_k^1 in the following sense.

The set

$$\{(d_2,\vec{\beta}) \; ; \; \vec{\beta} \text{ are associates for } \vec{\psi} \text{ and } \{d_2\}(\vec{\psi},\phi)$$
$$\text{is an essential subcomputation of } \{d\}(x,\phi) \text{ and}$$
$$\text{if } \psi_i = \varphi_n \text{ then } \beta_i = \gamma_n\}$$

is Σ_k^1 .

Proof: If we let β be a Δ_2^0-associate for ϕ we notice that this set is definable by a positive Σ_1^1-inductive definition relative to β and $(As(0),\ldots,As(k-1))$. The last predicate is Π_{k-2}^1 so the resulting set is Σ_{k-1}^1 (confer with the proof of Claim 3).

□ Claim 5

We define $D = \{e; \phi \text{ is used non-effectively at } e\}$ as before and see that it will be a Σ_{k-1}^1 subset of B . D will be finite and we may proceed as in the case $k = 1$.

This ends the proof of Theorem 6.27.

□

Remark 6.28

In this argument we showed that

$$1\text{-sc}(\phi) \subseteq A \subseteq \mu\text{-1-sc}(\phi)$$

so for the functionals constructed here there is no difference between the sets computable in ϕ and the sets μ-recursive in ϕ . This shows that asking for 1-sections and 2-envelopes does not reveal anything about the nature of Kleene-computations in general. These results are results about the countable functionals and about the power of S8.

The following corollary is due to J. Bergstra [1] but with a completely different proof working for $k = 2$ as well.

Corollary 6.29

Let $k > 2$. There is a functional ϕ of type k+1 with a top--less k-section, i.e. the k-section contains no maximal degree.

<u>Proof</u>: Pick $A \in \pi^1_{k-1} \smallsetminus \Sigma^1_{k-1}$ such that A is generated by its r.e. degrees and pick $\phi \in Ct(k+1)$ such that

$$A = 1-sc(\phi)$$

If for some $\psi \in Ct(k)$, $A = 1-sc(\psi)$, then A will be $\pi^1_{k-2}(h)$ for some $h \in A$. But $A \subseteq \Delta^0_2$ so A will be π^1_{k-2}, contradicting the choice of A. So ψ cannot be Kleene-equivalent to ϕ.

Moreover if $\psi \in k-sc(\phi)$ then $1-sc(\psi) \subseteq A$.
Since $1-sc(\psi) \in \pi^1_{k-2}$ there must be an $h \in A \smallsetminus 1-sc(\psi)$ and there will be a $\psi' \in k-sc(\phi)$ Kleene-equivalent to h. Then ψ' is not computable in ψ. This shows that ψ is not a maximal element in $k-sc(\phi)$ so $k-sc(\phi)$ is topless.
□

Recall that 'recursive' means 'countable recursive'.

<u>Corollary 6.30</u>
Let $\phi \in Ct(k)$, $A = \{f; f \text{ is recursive in } \phi\}$.
Then there is a $\Psi \in Ct(k+2)$ such that

$$A = 1-sc(\Psi)$$

<u>Proof</u>: By Corollary 5.53 there is an $h \in A$ such that

<u>i</u> A is generated by its r.e. degrees modulo h

<u>ii</u> A is $\pi^1_k(h)$

<u>iii</u> A is closed under recursion in finite lists.

We then use Theorem 6.27.
□

6.4. Another type-structure

Our main investigations have up to now dealt with either the maximal type-structure, the countable functionals or general type-structures In this section we will take a quick look at another specific type-structure. We will also look at a few concepts for type-structures in general.

<u>Definition 6.31</u>
Let $A = \langle A_k \rangle_{k \in \omega}$ be a type-structure closed under computations.

<u>a</u> We say that A is <u>weakly continuous</u> if ${}^2E\!\restriction\! A_1 \in A_2$

<u>b</u> We define the associates in A by

$As(1,A) = A_1$

$\alpha \in As(k+1),A)$ if

   <u>i</u> $\forall n\ Con(k+1,\bar{\alpha}(n))$

   <u>ii</u> α secures all elements in $As(k,A)$

<u>c</u> A is <u>strongly continuous</u> if for all $k \geq 1$

   <u>i</u> Each $\alpha \in As(k,A)$ is an associate for a $\varphi \in A_k$

   <u>ii</u> Each $\varphi \in A_k$ has an associate in $As(k,A)$

Remark 6.32

If A is strongly continuous then A is completely determined by A_1. The concepts of weakly continuous and strongly continuous are not the same.

Lemma 6.33

<u>a</u> If A is strongly continuous then A is weakly continuous.

<u>b</u> There is an A such that A is weakly continuous but not strongly continuous.

Proof:

<u>a</u> is trivial. If A_1 is nonempty and closed under recursion then there is no associate for ${}^2E\!\restriction\! A$.

<u>b</u> We define A_k by induction on k. Let F be the functional of type-2 constructed in Theorem 4.21 i.e. $F\!\restriction\! REC$ is not computable but $1\text{-sc}(F) = REC$. Let $A_0 = \omega$,

$$A_{k+1} = \{\psi : A_k \to \omega \, ; \, \psi \text{ is computable in } F\!\restriction\! REC\}.$$

Clearly A will be closed under computations, ${}^2E\!\restriction\! REC$ is not in A_2. $F\!\restriction\! REC \in A_2$ but $F\!\restriction\! REC$ has no recursive associate, since F would then be computable on REC.

□

Clearly if A_1 is the set of reals in some model of set theory all arguments from the previous sections are valid in A. But when we showed that $\pi_k^1 \subseteq 2\text{-en}({}^{k+2}0)$ and that every π_k^1-set of r.e. degrees generate a 1-section of a continuous type $k+2$ functional, we only used elementary properties about π_k^1 ; such as closure under number quantifications, the existence of a universal π_k^1-set etc.

We will now state without proof sufficient conditions for these results to hold. We will use some concepts from set theory and admissibility theory not defined in this book.

Definition 6.34

Let $B \subseteq Tp(1)$ and assume that B is closed under recursion and ordinary jump.

<u>a</u>   Let x be hereditarily countable, $f \in Tp(1)$. We say that <u>f is a code for x</u> if $\{(a,b); f(<a,b>) = 0\}$ is isomorphic to $<TC(x), \in>$ (TC = Transitive Closure).

<u>b</u>   By <u>the structure of B</u> we mean

$$Str(B) = \{x; \exists f \in B \ (f \text{ is a code for } x)\}$$

<u>c</u>   We say that B is <u>Σ_k-admissible</u> if $Str(B)$ is Σ_k-admissible (Σ_1-admissible = ordinary admissible).

Theorem 6.35

Let $A = <A_k>_{k \in \omega}$ be a strongly continuous type-structure. Assume that A_1 is closed under ordinary jump and assume that A_1 is Σ_k^1-admissible. Then

<u>a</u>   If $B \subseteq A_1$ is $(\pi_k^1)_{A_1}$ then B is semicomputable in $(^{k+2}0)_{A_{k+2}}$.

<u>b</u>   If $B \subseteq A_1$ is $(\pi_k^1)_{A_1}$, B is closed under recursion in finite lists and B is recursively generated from its r.e. elements then there is a functional $\psi \in A_{k+2}$ such that inside A

$$B = 1\text{-sc}(\psi) = \mu\text{-1-sc}(\psi)$$

Remark 6.36

<u>a</u>   In the proofs of Corollary 5.33 and Theorem 6.27 we never used properties about π_k^1-sets that do not follow easily from Σ_k-admissibility. The recursion-theoretic part of the proof of this theorem will just be a repetition of previous arguments and there is no need to give them here.

<u>b</u>   The equation $2\text{-en}(^{k+2}0) = \pi_k^1$ is false in general for type-structures. Our only application of Theorem 6.35 will give a counter-

example. Notice that $\pi_1^1 \subseteq$ 1-en(ψ) for all functionals ψ of type ≥ 2 within all type-structures.

Corollary 6.37

Let $B \subseteq Tp(1)$ be closed under recursion in finite lists. Assume that B is recursively generated from its r.e. degrees and that B is Σ_1^1.

Then there is a functional $\psi \in Tp(3)$ (in the maximal type structure) such that $B = $ 1-sc(ψ).

Proof: Let us regard the hyperarithmetical type-structure HAT defined as follows:

$HAT_0 = \omega$

$HAT_1 = \Delta_1^1 = $ the class of hyperarithmetical functions

$HAT_{k+1} = \{\psi: HAT_k \to \omega;\ \psi$ has a hyperarithmetical associate$\}$

HAT is a strongly continuous type-structure and HAT_1 is well known to be Σ_1-admissible.

By the well-known Spector-Gandy theorem (see e.g. Rogers [38]) we know that for $C \subseteq \omega$ we have

C is Σ_1^1 if and only if C is $(\pi_1^1)_{\Delta_1^1}$ i.e. C is $(\pi_1^1)_{HAT_1}$

Since B is determined by $\{e;\ W_e \in B\}$, B will be $(\pi_1^1)_{HAT_1}$ and by Theorem 6.35 $\underline{b}$ there is a $\psi_0 \in HAT_3$ such that inside HAT

$$1-sc(\psi_0) = B$$

But by Theorem 2.6 there is a $\psi \in Tp(3)$ such that

$$1-sc(\psi) = (1-sc(\psi_0))_{HAT} = B$$

□

Sacks' plus-one theorem [40] and [41] is one of the beautiful regularity-results of recursion in normal functionals. It says that if ϕ is a normal functional of type ≥ 2 then there is a normal functional $F \in Tp(2)$ such that

$$1-sc(F) = 1-sc(\phi)$$

A consequence of Corollary 6.37 is that this result is as false as possible if we look at recursion in an arbitrary functional.

Corollary 6.38

There is a functional $\psi \in Tp(3)$ such that for all $F \in Tp(2)$

$$1\text{-sc}(\psi) \neq 1\text{-sc}(F)$$

Proof: Choose B in Corollary 6.37 such that $B \in \Sigma_1^1 \smallsetminus \Pi_1^1$ and choose $\psi \in Tp(3)$ such that $B = 1\text{-sc}(\psi)$.

If $F \in Tp(2)$ such that $1\text{-sc}(F) \subseteq B$ then F cannot be normal so by Corollary 6.19 we may assume that F is continuous. Then $1\text{-sc}(F)$ is $\Pi_1^1(h)$ for some $h \in 1\text{-sc}(F)$, so $1\text{-sc}(F)$ is a proper subset of $1\text{-sc}(\psi)$.

□

Problems 6.39

We know that within the countable functionals new 1-sections crop up as we move up in type, except from two to three where the problem is open.

We do not know if this is the case in the maximal type-structure. We formulate two open problems as conjectures below. We have no mathematical reason for suggesting these solutions to the problems, they just seem to be the most likely ones.

Conjectures

a   There is a $\psi \in Ct(3)$ such that for all $F \in Tp(2)$

$$1\text{-sc}(F) \neq 1\text{-sc}(\psi)$$

b   For each $k \geq 1$ there is a $\psi \in Tp(k+1)$ such that for all $\varphi \in Tp(k)$

$$1\text{-sc}(\psi) \neq 1\text{-sc}(\varphi)$$

7. SOME FURTHER RESULTS AND TOPICS

7.1 Irreducible and nonobtainable functionals

This section will be a continuation of chapter 4. In section 4.2 we constructed irreducible functionals of type-2 and in section 6.3 we constructed irreducible functionals of arbitrary types > 3. The first construction of an irreducible functional was due to Hinman [17]. He constructed an $F \in Ct(2)$ which is not recursively equivalent to any f. Our construction in 6.3 only works for Kleene-computations. Later Sergey Dvornickov [6] improved these results. Theorem 7.1 with corollaries are due to him. In his proof he used the f-space interpretation, see Ershov [7], [8] and [9].

Theorem 7.1

Let $k \geq 2$. Let $V_n = \{\varphi \in Ct(k-1); \varphi(^{k-2}0) \leq n\}$. Let $e, d, \psi \in Ct(k)$, $\varphi \in Ct(k-1)$, $n \in \omega$ be given such that ψ is recursive in φ via e and φ is recursive in ψ via d. Let $\psi_n = \psi \upharpoonright V_n$.

Then there is a $\psi' \in Ct(k)$ extending ψ_n and a $m > n$ such that for no ψ'' extending ψ'_m and no $\varphi' \in Ct(k-1)$

ψ'' is recursive in φ' via e and φ' is recursive in ψ'' via d.

Proof:
Let π_e and π_d be the recursive maps with indices e, d resp. such that

$$\beta \in As(\varphi) \rightarrow \pi_e(\beta) \in As(\psi) \wedge \alpha \in As(\psi) \rightarrow \pi_d(\alpha) \in As(\varphi)$$

Let $A = \{\pi_e(\beta); \beta \in As(\varphi)\}$. Then $A \subseteq As(\psi)$ is $\underset{\sim}{\Sigma}^1_{k-1}$.

Fix σ_0 such that $B^{k-1}_{\sigma_0}$ contains more than one element and such that

$$\varphi \in B^{k-1}_{\sigma_0} \rightarrow \varphi(^{k-2}0) = n+1$$

By Lemma 5.31 there is an associate α for ψ such that

$$\forall \beta \in As(\varphi) \exists \sigma (\sigma_0 \prec \sigma \wedge Con(k-1, \sigma) \wedge \alpha(\sigma) = 0 \wedge \pi_e(\beta)(\sigma) > 0)$$

By inspection of the proof of Lemma 5.31 we may assume that there is no finite set $\tau_1, \ldots, \tau_t$ such that $\alpha(\tau_1) = \ldots = \alpha(\tau_n) > 0$ and

$$B^{k-1}_{\sigma_1} \subseteq B^{k-1}_{\tau_1} \cup \ldots \cup B^{k-1}_{\tau_n}$$

See also Lemma 5.10.

Moreover we see that $B_\sigma^{k-1} \subseteq V_{n+1} \smallsetminus V_n$.

Let $\beta_0 = \pi_d(\alpha)$. Then

$$\exists \sigma (\mathrm{Con}(k-1,\sigma) \wedge B_\sigma^{k-1} \subseteq V_{n+1} \smallsetminus V_n \wedge \alpha(\sigma) = 0 \wedge \pi_e(\pi_d(\alpha))(\sigma) > 0)$$

Let t be so large that $\pi_e(\pi_d(\alpha))(\sigma)$ is computable from $\bar{\alpha}(t)$. Let α' be an extension of $\bar{\alpha}(t)$ to an associate for ψ' such that

<u>i</u> $\varphi \notin V_{n+1} \smallsetminus V_n \rightarrow \psi'(\varphi) = \psi(\varphi)$

<u>ii</u> ψ' is not constant on B_σ^{k-1}

Then $\pi_e(\pi_d(\alpha'))(\sigma) > 0$ so $\pi_e(\pi_d(\sigma'))$ cannot be an associate for ψ' . But then ψ' cannot be recursively equivalent to any φ' via e,d .

Let m_0 be maximal such that $\bar{\alpha}(t)$ contains information about ψ_{m_0} (and then also about ψ'_{m_0}). Let $m = \max\{n+1, m_0\}$. Then any functional ψ'' extending ψ'_m will have an associate extending $\bar{\alpha}(t)$ but will not be constant on B_σ^{k-1} . The argument above then shows that ψ'' is not reducible via e,d .   □

Corollary 7.2

Let $k \geq 2$. Then there is a $\psi \in Ct(k)$ that is not recursively equivalent to any $\varphi \in Ct(k-1)$.

<u>Proof</u>: Using Theorem 7.1 we construct an increasing sequence ψ_n on sets V_{m_n} such that for $n = \langle e,d \rangle$, no extension of ψ_n to $Ct(k-1)$ is reducible via e,d . If we let $\psi = \underset{n \in \omega}{\cup} \psi_n$ we see that ψ is recursively irreducible.   □

Remark 7.3

<u>a</u>   As a special case we get that ψ is not Kleene-reducible.

<u>b</u>   By inspection of the proof one will see that ψ may actually be computable in $0'$.

<u>c</u>   As a corollary to the proof of Theorem 7.1 we see that we may produce a set $A \subseteq Ct(k)$ of the same power as the continuum such that no $\psi \in A$ is recursively reducible. A may be chosen as a compact subset of $Ct(k)$.

When we work with recursion Corollary 7.2 tells us that there are new degrees in all types. This is the best we could hope for since all

degrees will be bounded by the degree of a function since all functionals are recursive in any of its associates.

If we discuss Kleene-degrees the situation is different. We know
that the degrees of the functions are not dense in the set of all degrees, the fan-functional ϕ is a counterexample, Γ another (see
section 4.3 and 4.4). In this section we will construct non-obtainable
functionals of arbitrary high type showing that the degrees of the
type-k-functionals are not dense in the set of all degrees.

First we need to build up some machinery. We will use the notation
from section 5.2.

Lemma 7.4

Let $k \geq 3$ and let $\vec{\psi}$ be a finite sequence of functionals of
type $\leq k$. Assume that $\{e_0\}(\vec{\psi})$ is a convergent computation. Then
there is a set $A \subseteq H_{k-1}$ such that

<u>i</u> A is $\underset{\sim}{\Sigma}_{k-2}$

<u>ii</u> If $\psi_i \in Ct(k)$, $\varphi \in Ct(k-1)$ and $\psi_i(\varphi)$ occurs in a subcomput-
ation of $\{e_0\}(\vec{\psi})$ then $h_\varphi \in A$.

Proof: Let $\vec{\alpha}$ be associates for $\vec{\psi}$ resp. We know that in all sub-
computations $\{e\}(\vec{\psi})$ of $\{e_0\}(\vec{\psi})$ all arguments in $\vec{\psi} \smallsetminus \vec{\psi}$ will be of
type $\leq k-2$.

Let

$\quad\quad C = \{<e,\vec{\gamma},s>; \ \vec{\gamma}$ are associates for functionals $\vec{\psi}$ such that
$\quad\quad\quad\quad\quad\quad \{e\}(\vec{\psi}) \simeq s$ is a subcomputation of $\{e_0\}(\psi)$, and
$\quad\quad\quad\quad\quad\quad$ if $\psi_i = \psi_j$ of type k or $k-1$ then $\gamma_i = \alpha_j\}$.

It is easy to see that C is $\Sigma^1_{k-2}(\vec{\alpha})$.

Now we use f from Theorem 2.28. We let $h \in A$ if there is
$<e,\vec{\gamma},s>$ in C such that e is an index for

$$\psi_j(\lambda\xi\{e_1\}(\xi,\vec{\psi})) \ ,$$

ψ_j is of type k and

$$\forall n \ \exists m \ f(<e_1,\bar{\beta}_n(m) \ , \bar{\gamma}_1(m),\ldots,\bar{\gamma}_p(m)>) = h(n)+1$$

where p is the length of $\vec{\gamma}$ and β_n is a canonical associate for
φ_n^{k-2} .

A is clearly $\underset{\sim}{\Sigma}^1_{k-2}(\vec{\alpha})$ and has the property we want.

□

Definition 7.5

Let $k \geq 1$, $\psi \in Ct(k)$. We call α a __semi-associate__ for ψ if $\forall n \; \psi \in B^k_{\bar{\alpha}(n)}$. A number is its own semi-associate.

Remark 7.6

__a__ A function f will be its own semi-associate.

__b__ If $k \geq 2$ then α is a semi-associate for $\psi \in Ct(k)$ if and only if

__i__ $\forall n \; Con(k, \bar{\alpha}(n))$

__ii__ $\forall \sigma, t \; (\alpha(\sigma) = t+1 \rightarrow \psi$ is constant t on $B^{k-1}_\sigma)$

The following lemma is a generalization of Theorem 4.54.

Lemma 7.7

Let f be as in Theorem 2.28.
Let $\vec{\psi} = (\psi_1, \ldots, \psi_k)$ and let $\vec{\alpha}$ be semi-associates for $\vec{\psi}$ resp. Assume that $\{e\}(\vec{\psi}) \simeq s$ and that whenever $\psi_j(\varphi)$ is used in a sub-computation of $\{e\}(\vec{\psi})$ then α_j secures all associates for φ $(j \leq k)$. Then

$$\exists n \; f(<e, \bar{\alpha}_1(n), \ldots, \bar{\alpha}_k(n)>) = s+1 \; .$$

__Proof:__ The proof is as in Theorem 2.28 and we leave it for the reader.
$\square$

In the construction of a non-obtainable functional we need a rather special dense subset of $Ct(k-1)$.

Lemma 7.8

Let $k \geq 2$. There is a primitive recursive family $\{\xi_\sigma; Con(k-1, \sigma)\}$ of functionals in $Ct(k-1)$ satisfying

__i__ $\xi_\sigma \in B^{k-1}_\sigma$

__ii__ If $\sigma_1 < \sigma_2$ and $B^{k-1}_{\sigma_2} \not\subseteq B^{k-1}_{\sigma_1}$ then $\xi_{\sigma_2} \in B^k_{\sigma_1}$

__iii__ If $\sigma_1 \neq \sigma_2$ then either $\xi_{\sigma_1} \neq \xi_{\sigma_2}$ or $B^{k-1}_{\sigma_1} = B^{k-1}_{\sigma_2} = \{\xi_{\sigma_1}\}$

__Proof:__ We construct ξ_σ by primitive recursion on σ. So assume that ξ_τ is defined for all $\tau < \sigma$. Let $\tau_1, \ldots, \tau_n$ be those $\tau < \sigma$ such that $B^{k-1}_\sigma \not\subseteq B^{k-1}_\tau$. By Lemma 5.10 __c__ $B^{k-1}_\sigma \not\subseteq \underset{i \leq n}{\cup} B^{k-1}_{\tau_i}$.

By Corollary 5.11 __c__ we may find σ' extending σ such that

$$B_{\sigma'}^{k-1} \cap \bigcup_{i \leq n} B_{\tau}^{k-1} = \emptyset$$

If B_{σ}^{k-1} contains more than one element it follows from the proof of
5.11 $\underline{c}$ that we may pick σ' such that $B_{\sigma'}^{k-1}$ contains more than one
element.
Let $\xi_{\sigma} = \text{Ext}^{k-1}(t,\sigma')$ where t is larger than any number having
entered into the construction so far. (See 5.5 and 5.7 for Ext^{k-1}.)
By construction $\underline{i}$ and $\underline{ii}$ will hold.

Claim
If $\text{Ext}^{t}(t,\sigma')$ never takes the value t, then $B_{\sigma'}^{k-1}$ contains
just one element.

Proof: For $k = 2$ or $k = 3$ the claim is easy and for $k > 3$ we leave
the claim as an exercise for the reader.
$\quad\quad\quad\quad\quad\quad\quad\quad\quad\quad\quad\quad\quad$ □ Claim

$\underline{iii}$ of the lemma follows directly from the claim and the construction.
$\quad\quad\quad\quad\quad\quad\quad\quad\quad\quad\quad\quad\quad\quad\quad\quad\quad\quad\quad\quad\quad\quad\quad\quad\quad\quad\quad\quad$ □

Remark 7.9
Throughout this section we will let $\{\xi_{\sigma} ; \text{Con}(k-1,\sigma)\}$ be as in
Lemma 7.8. We will also use the following property, a consequence of
Lemma 5.12:
The relation ' $\xi \in B_{\tau}^{k-1}$ ' is primitive recursive.

Definition 7.10
Let $k \geq 3$. For each $\psi \in \text{Ct}(k)$ define the sequence δ_{m}^{ψ} of
length m as follows:
For $\sigma < m$ let

$$\delta_{m}^{\psi}(\sigma) = \begin{cases} t+1 & \text{if } \exists \pi < m \ (\sigma \nleqq \pi) \\ & \quad \wedge \ \forall \pi < m \ (\sigma \leqq \pi) \Rightarrow \psi(\xi_{\pi}) = t \\ 0 & \text{otherwise} \end{cases}$$

where we assume $\text{Con}(k-1,\sigma)$ and $\text{Con}(k-1,\pi)$.

Remark 7.11
Using a definition like this we may get an alternative proof of
Corollary 5.18. We will not in general have $\text{Con}(k,\delta_{m}^{\psi})$ but with minor
changes in the definition this could be arranged. δ_{m}^{ψ} is designed for
other purposes than giving a proof of 5.18. Still the next lemma is
trivial and we leave the proof for the reader.

Lemma 7.12

$\lim_{m \to \infty} \delta_m^\psi$ is the principal associate for ψ . □

Lemma 7.13

Let $k \geq 3$ and let K be a compact set such that

i $\forall \beta \in K \; \forall n \; \text{Con}(k, \bar{\beta}(n))$

ii $\forall \beta \in K \; (\beta \notin As(k))$

Let

$$\Delta_K(\psi) = \mu n \; \forall m \geq n \; \forall \beta \in K \; \exists \sigma < m \, (\text{Con}(k-1, \sigma) \land \beta(\sigma) = 0 \land \delta_m^\psi(\sigma) > 0)$$

Then Δ_K is total and has an associate recursive in any suitable description of K as a compact set, e.g. in

$$\{<n, \tau_1, \ldots, \tau_s>; \{\tau_1, \ldots, \tau_s\} = \{\tau; \text{lh}(\tau) = n \land B_\tau^1 \cap K \neq \emptyset\}\}$$

Proof: Let $\psi \in Ct(k)$ and let α be an associate for ψ . We will show how to compute $\Delta_K(\psi)$ from α and K . Let $\beta \in K$. If

$$\forall \sigma \; (\text{Con}(k-1, \sigma) \land \alpha(\sigma) > 0 \rightarrow \beta(\sigma) > 0)$$

then $\beta \in As(k)$ since $\forall n \; \text{Con}(k, \bar{\beta}(n))$.
This is not the case so

$$\forall \beta \in K \; \exists \sigma \; (\text{Con}(k-1, \sigma) \land \alpha(\sigma) > 0 \land \beta(\sigma) = 0)$$

Since K is compact there is a finite set $\{\sigma_1, \ldots, \sigma_s\}$ such that $\forall i \leq s \; \text{Con}(k-1, \sigma_i)$ and

$$\forall \beta \in K \; \exists i \leq s \; (\alpha(\sigma_i) > 0 \land \beta(\sigma_i) = 0)$$

Let m_0 be so large that for all $i \leq s$ σ_i has a proper extension $\pi_i < m$ such that $\text{Con}(k-1, \pi_i)$.
Then

$$\forall \beta \in K \; \forall m \geq m_0 \; \exists i \; (\delta_m^\psi(\sigma_i) > 0 \land \beta(\sigma_i) = 0)$$

We may find such m_0 and $\{\sigma_1, \ldots, \sigma_s\}$ uniformly recursive in α, K .
Now

$$\Delta_K(\psi) = \mu n \; \forall m \; (n \leq m \leq m_0 \rightarrow \forall \beta \in K \; \exists \sigma \; (\text{Con}(k-1, \sigma) \land \delta_m^\psi(\sigma) > 0 \land \beta(\sigma) = 0))$$

which is uniformly recursive in m_0, K . But this means that $\Delta_K(\psi)$ is uniformly recursive in K and any associate for ψ , so Δ_K will have an associate recursive in K . □

We will now produce a compact set K such that Δ_K is recursive but not computable in any $\psi \in Ct(k)$. It is in this proof we will make use of the first line in the definition of δ_m^ψ. In the proof of Lemma 7.13 this line was just a complicating factor, and so it also is in the proof of Lemma 7.12.

Definition 7.14

Let $k \geq 3$. Define the relation Σ by

$$\Sigma(\alpha, h) \leftrightarrow \exists B \ (\ B \text{ is } \Sigma_{k-2}^1(\alpha) \wedge B \subseteq H_{k-1} \wedge h \in B)$$

For each α let $\Sigma_\alpha = \{h; \Sigma(\alpha, h)\}$.

Lemma 7.15

__a__ Σ is Π_{k-1}^1

__b__ Each Σ_α is a proper subset of H_{k-1}

__c__ If $B \subseteq H_{k-1}$ is $\underset{\sim}{\Sigma}_{k-2}^1$ then there is an $\alpha \in \{0,1\}^{\mathbb{N}}$ such that $B \subseteq \Sigma_\alpha$.

Proof:

__a__ is trivial.

__b__ Each Σ_α will be a $\underset{\sim}{\Sigma}_{k-2}^1$-subset of H_{k-1}. By Lemma 5.28 Σ_α will then be a proper subset of H_{k-1}.

__c__ is trivial from the definition of Σ.

$\square$

By Corollary 5.23 there is a primitive recursive relation R such that

$$\Sigma(\alpha, h_1) \leftrightarrow \forall h_2 \in H_{k-1} \ \exists n \ R(\bar{\alpha}(n), \bar{h}_1(n), \bar{h}_2(n), n)$$

We use this relation R in the next definition.

Definition 7.16

Let $k \geq 3$.

__a__ Assume $Con(k-1, \sigma)$.

Let

$$\sigma_i(\delta) = \begin{cases} (\sigma(\delta)-1)_i +1 & \text{if } \sigma(\delta) > 0 \\ 0 & \text{if } \sigma(\delta) = 0 \end{cases} \quad i = 1,2$$

where $(\ )_1$ and $(\ )_2$ are the two projection maps of $<,>$.

<u>b</u>   Assume Con(k-1,σ) , and assume that B_σ^{k-1} has more than one
element.

Let h_σ be the largest sequence $\langle h_\sigma(0),...,h_\sigma(n)\rangle$ such that

$$h_\sigma(t) = s \quad \text{if } \exists\delta \ (\sigma(\delta) = s+1 \wedge \varphi_t^{k-2} \in B_\delta^{k-2})$$

(By the assumption on σ there will be a t such that $h_\sigma(t)$ is
undefined, and we then see that $h_\sigma(t')$ will be undefined for t'>t)

<u>c</u>   Let

$$P_\alpha(\sigma) = \begin{cases} 1 & \text{if } \exists n \ R(\bar{\alpha}(n),\bar{h}_{\sigma_1}(n),\bar{h}_{\sigma_2}(n),n) \\ & \text{or if } B_\sigma^{k-1} \text{ contains just one element} \\ \\ 0 & \text{otherwise} \end{cases}$$

<u>d</u>   Let $K_k = \{P_\alpha; \alpha \in \{0,1\}^{\mathbb{N}}\}$.

Lemma 7.17

Let $k \geq 3$, P_α be defined as above.

<u>a</u> P_α is a semi-associate for $^k 0$.

<u>b</u> P_α is not an associate.

<u>c</u>   If $B \subseteq H_{k-1}$ is $\underset{\sim}{\Sigma}_{k-2}^1$ then there is a $P_\alpha \in K_k$ securing all
associates for ξ whenever $h_\xi \in B$.

<u>d</u>   If $P_\alpha(\sigma) = 1$ and $B_\tau^{k-1} \subseteq B_\sigma^{k-1}$ then $P_\alpha(\tau) = 1$.

Proof:

<u>a</u>   is trivial.

<u>b</u>   Let α be given. By Lemma 7.15 <u>b</u> there is an $h_1 \in H_{k-1} \smallsetminus \Sigma_\alpha$.
Then $\neg\Sigma(\alpha,h_1)$. Let $h_2 \in H_{k-1}$ be such that
$\forall n \ \neg R(\bar{\alpha}(n),\bar{h}_1(n),\bar{h}_2(n))$. Let $h_1 = h_{\xi_1}$, $h_2 = h_{\xi_2}$ and let $\xi = \langle\xi_1,\xi_2\rangle$.
It is easy to see that P_α will not secure any associate for ξ .

<u>c</u>   For $h \in B$ let $h_1(n) = (h(n))_1$ and $h_2(n) = (h(n))_2$. Let
$B_1 = \{h_1; h \in B\}$. By Lemma 7.15 <u>c</u> there is an $\alpha \in \{0,1\}^{\mathbb{N}}$ such
that $B_1 \subseteq \Sigma_\alpha$, so

$$h \in B \leftrightarrow \exists n \ R(\bar{\alpha}(n),\bar{h}_1(n),\bar{h}_2(n),n) .$$

It is easy to see that $h_\xi \in B \rightarrow P_\alpha$ secures all associates for ξ .

<u>d</u>   If $B_\tau^{k-1} \subseteq B_\sigma^{k-1}$ then $h_\sigma \prec h_\tau$, $B_{\tau_1}^{k-1} \subseteq B_{\sigma_1}^{k-1}$ and $B_{\tau_2}^{k-1} \subseteq B_{\sigma_2}^{k-1}$.

This is a trivial observation and $\underline{d}$ follows trivially from this observation. For the first part we may use Lemma 5.10 $\underline{b}$.

$\square$

The main result of this section will be

Theorem 7.18

Let $k \geq 3$ and let K_k and $\Delta = \Delta_{K_k}$ be defined as above.

$\underline{a}$ $\Delta \in Ct(k+1)$ and has a recursive associate.

$\underline{b}$ Δ is not computable in any $\psi \in Ct(k)$.

Proof:

$\underline{a}$ h_σ is uniformly recursive in σ , so P_α is uniformly recursive in α . It follows that K_k is recursively compact so Δ has a recursive associate by Lemma 7.13.

$\underline{b}$ Assume that this is not the case. Then there is a $\psi \in Ct(k)$ and an index e such that

$$\forall \varphi \in Ct(k)(\Delta(\varphi) = \{e\}(\varphi, \psi))$$

We will look at $\{e\}(^k 0, \psi)$. By Lemma 7.4 there is a $\underset{\sim}{\Sigma}^1_{k-2}$-set $B \subseteq H_{k-1}$ such that if $^k 0(\xi)$ is used in a subcomputation of $\{e\}(^k 0, \psi)$ then $h_\xi \in B$. By Lemma 7.17 $\underline{c}$ there is a $P_\alpha \in K_k$ such that whenever $h_\xi \in B$ then P_α secures all associates for ξ . By Lemma 7.7 there is a number n such that if $\varphi \in B^k_{P_\alpha}(n)$ then

$$\Delta(\varphi) = \{e\}(\varphi, \psi) = \{e\}(^k 0, \psi) = \Delta(^k 0)$$

We defined δ^φ_m for $\varphi \in Ct(k)$ but the definition makes sense for all φ defined on all ξ_σ . Define φ_0 by

$$\varphi_0(\xi_\sigma) = \begin{cases} 0 & \text{if } B^{k-1}_\sigma \text{ contains just one element or if} \\ & \quad \exists \tau < n \ (\xi_\sigma \in B^{k-1}_\tau \wedge P_\alpha(\tau) = 1) \\ \sigma+1 & \text{otherwise} \end{cases}$$

Claim 1

$\underline{a}$ φ_0 is well-defined on $\{\xi_\sigma; Con(k-1,\sigma)\}$

$\underline{b}$ For any finite set $\{\sigma_1,\ldots,\sigma_t\}$ there is a $\varphi \in B^k_{P_\alpha}(n)$ such that $\varphi(\xi_{\sigma_1}) = \varphi_0(\xi_{\sigma_1}) \wedge \ldots \wedge \varphi(\xi_{\sigma_t}) = \varphi_0(\xi_{\sigma_t})$.

Proof:

<u>a</u>   is trivial from Lemma 7.8 <u>iii</u>

<u>b</u>   Pick $\varphi' \in B^k_{P_\alpha}(n)$. Let $X = U\{B^{k-1}_\tau ; \tau < n \wedge P_\alpha(\tau) = 1\}$

For each i such that $\xi_{\sigma_i} \notin X$ let X_i be a closed-open neigh-bourhood of ξ_{σ_i} such that $X_i \cap X = \emptyset$. We may assume $i \neq j \Rightarrow X_i \cap X_j = \emptyset$.
Let

$$\varphi(\xi) = \begin{cases} \sigma_i + 1 & \text{if } \xi \in X_i \text{ and } \xi_{\sigma_i} \notin X \\ \varphi'(\xi) & \text{otherwise} \end{cases}$$

It is not hard to see that ξ has the desired properties.
                                                                                    □ Claim 1

<u>Claim 2</u>

<u>a</u> $\forall m > n \; \forall \sigma \; (n \leq \sigma \leq m \Rightarrow \delta^{\varphi_0}_m(\sigma) \leq P_\alpha(\sigma))$ where we assume Con(k-1,σ)

<u>b</u>   If $\sigma < n$, Con(k-1,σ) and $P_\alpha(\sigma) = 0$ then there is an m_0 such
      that

$$m > m_0 \Rightarrow \delta^{\varphi_0}_m(\sigma) = 0$$

Proof: In this proof we will always assume Con(k-1,σ) . For each
m,σ we have that $\delta^{\varphi_0}_m(\sigma)$ is either 0 or $\varphi_0(\xi_\sigma)+1$ and $\varphi_0(\xi_\sigma)$ is
either 0 or $\sigma+1$.
If $\delta^{\varphi_0}_m(\sigma) = \sigma+2$ then $\exists \sigma_1 < m \; (\sigma \nleq \sigma_1 \wedge \varphi_0(\xi_\sigma) = \varphi_0(\xi_{\sigma_1}) = \sigma+1)$.
But when $\sigma_1 \neq \sigma$ this is impossible. So $\delta^{\varphi_0}_m(\sigma) \in \{0,1\}$.

<u>a</u>   Assume that $n \leq \sigma < m$ and $\delta^{\varphi_0}_m(\sigma) = 1$.
      Since for all φ

$$\delta^\varphi_m(\sigma) > 0 \Rightarrow \delta^\varphi_m(\sigma) = \varphi(\xi_\sigma)+1$$

we must have $\varphi_0(\xi_\sigma) = 0$. If this is because B^{k-1}_σ contains just
one element we have constructed P_α in such a way that $P_\alpha(\sigma) = 1$.
      If B^{k-1}_σ contains more than one element we must have

$$\exists \tau < n \; (\xi_\sigma \in B^{k-1}_\tau \wedge P_\alpha(\tau) = 1)$$

Then $\tau < \sigma$ and since $\xi_\sigma \in B^{k-1}_\tau$ we see by Lemma 7.8 <u>ii</u> that $B^{k-1}_\sigma \subseteq B^{k-1}_\tau$
By Lemma 7.17 <u>d</u> then $P_\alpha(\sigma) = 1$. So $\delta^{\varphi_0}_m(\sigma) = 1 \Rightarrow P_\alpha(\sigma) = 1$ and
$\delta^{\varphi_0}_m(\sigma) \leq P_\alpha(\sigma)$.

<u>b</u>   If $P_\alpha(\sigma) = 0$ then B^{k-1}_σ contains more than one element. If

$B_\sigma^{k-1} \subseteq \cup\{B_\tau^{k-1}; \tau < n \wedge P_\alpha(\tau) = 1\}$ then by Lemma 5.10 $\underline{c}$ there is a $\tau < n$ such that $B_\sigma^{k-1} \subseteq B_\tau^{k-1}$ and $P_\alpha(\tau) = 1$. But then $P_\alpha(\sigma) = 1$ by Lemma 7.17 $\underline{d}$, contradicting the assumption.

So $B_\tau^{k-1} \nsubseteq \cup\{B_\tau^{k-1}; \tau < n \wedge P_\alpha(\tau) = 1\}$. By Corollary 5.11 $\underline{c}$ there are extensions σ_1 and σ_2 of σ such that

$\sigma_1 \nleq \sigma_2$ and $B_{\sigma_1}^{k-1} \cap \cup\{B_\tau^{k-1}; \tau < n \wedge P_\alpha(\tau) = 1\} = \emptyset$.

W.l.o.g. we may assume that $B_{\sigma_2}^{k-1}$ contains more than one element. Then $\varphi_0(\xi_{\sigma_1}) = \sigma_1 + 1$ and $\varphi_0(\xi_{\sigma_2}) = \sigma_2 + 1$. For $m > \sigma_2$ we then see $\delta_m^{\varphi_0}(\sigma) = \emptyset$.

$$\square \ \text{Claim 2}$$

By Claim 2 we have

$$\exists m_0 \ \forall m \geq m_0 \ \forall \sigma < m \ (\text{Con}(k-1,\sigma) \Rightarrow \delta_m^{\varphi_0}(\sigma) \leq P_\alpha(\sigma)).$$

Choose $m > \max\{\Delta(^k0), m_0\}$. By Claim 1 $\underline{b}$ let $\varphi \in B_{\bar{P}_\alpha(n)}^k$ be such that

$$\forall \sigma < m \ (\text{Con}(k-1,\sigma) \Rightarrow \varphi(\xi_\sigma) = \varphi_0(\xi_\sigma))$$

Since $\delta_m^\varphi = \delta_m^{\varphi_0}$ we see $\Delta(\varphi) > m$. But by choice of n $\Delta(\varphi) = \Delta(^k0)$ since $\varphi \in B_{\bar{P}_\alpha(n)}^k$, contradicting $m > \Delta(^k0)$. This proves the Theorem.

$$\square$$

Corollary 7.19

If $k \geq 2$ then there is a recursive functional of type $k+1$ which is not computable in any functional of type k.

Proof: For $k = 2$ we may use the fan-functional as an example. For $k > 2$ we use Theorem 7.18.

Remark 7.20

It has later been shown that for each $k > 3$ and each $\Delta \in Ct(k)$ there is a $\Delta' \in Ct(k)$ such that Δ' is not computable in Δ and any $\varphi \in Ct(k-1)$.

7.2 Concluding remarks

In this book we have discussed some of the aspects of recursions and computability over the countable or continuous functionals. We have concentrated on a direct approach based on the original definitions of Kleene [23] and Kreisel [24]. There are important aspects left untouched, partly because they are not within the scope of the book, partly because they are well covered in the literature.

Some of them are however, so important that it would be misleading not to mention them.

Martin Hyland [19], [20] and [21] investigated structural approaches like the ones we gave in Chapter 3. He also derived results about the computable structure in an abstract setting, like the existence of a computably enumerated dense subset of $Ct(k)$. Several of our results in chapter 5 are in [20] or [21], or can easily be obtained from these papers.

We have concentrated on type-structures with total functionals and on partial maps on Cartesian products of types. But there are interesting results and problems concerning hereditarily partial functionals. This work was initiated in Platek [37], and Moldestad [27] gives a thorough survey of Platek's results. The connections between partial and total continuous functionals was investigated by Ershov, and in [7], [8] and [9] he gives important contributions to the structural theory of the continuous functionals. Feferman [10] discusses the recent development of this area of research. We recommend these papers as an introduction to a further study of partial functionals, and of the lattice-theoretic approach to the continuous functionals due to Ershov.

We have not given any applications of this theory. In his original paper Kreisel [24] discussed the connection with constructive analysis. Using $<Ct(k)>_{k\in\omega}$ we may give a constructive interpretation of formulas of analysis.

A growing area where this or similar theories may have applications is effective algebra. In Moldestad, Stoltenberg-Hansen and Tucker [29], [30] and in Tucker [44] the effective content of an algebraic structure is investigated and several interesting equivalence results and applications are obtained. This work was initiated by J.V. Tucker and we think that there are possibilities for mutual applications between the theory for computations in algebra and the theory for computations on the continuous functionals. Any reader interested in modern down-to-earth computation theory should consult these papers.

Our final remarks will be on the future problems in this area of mathematics.

We think that the structure of the continuous or countable functionals is well understood. Through Ershov [7], [8] and [9], Hyland [19], [20] and [21] and Chapters 3 and 5 of this book we are given a large variation of approaches to Ct(k) and any argument leaning on the structural properties of Ct(k) will find what it needs somewhere in these expositions.

We have formulated precise open problems elsewhere in the book. Here we will mention two unprecise but fundamental problems. That is really to "understand" computations and recursions. Though we have answered central problems concerning sections, envelopes, reducibility and obtainability we did not have to enter deeply into the complex world of computations. This is reflected in the fact that we could not characterize 1-sections of type-two functionals in a satisfactory way, probably because this really would involve an understanding of the mechanism of computations. Though the characterizations of sections and envelopes for higher type theories are interesting results in themselves, the arguments leading up to them do not seem to lead anywhere further. The main problem now is to find the right questions.

THE END

BIBLIOGRAPHY

[1]   Bergstra, J.A.: Computability and continuity in finite types,
      Thesis, University of Utrecht, 1976.

[2]   Bergstra, J.A.: 1-envelopes of Continuous Functionals,
      Recursive Function Theory: Newsletter No 16 (1977), Item 196.

[3]   Bergstra, J.A.: The continuous functionals and ^2E, in J.E.
      Fenstad, R.O. Gandy and G.E. Sacks (eds.), Generalized Recursion
      Theory II, North-Holland 1978.

[4]   Bergstra, J.A. and Wainer, S.S.: The 'real' ordinal of the
      1-section of a continuous functional. Talk contributed to Logic
      Colloquium '76, Oxford 1976, Abstract in J.S.L.

[5]   Cenzer, D.: Inductively defined sets of reals, Bulletin of A.M.S.
      80 (1974) 485-487.

[6]   Dvornickov, S.G.: On e-degrees of everywhere defined functionals
      (in russian), Logica i Algebra No 18 (1979), 32-46.

[7]   Ershov, Yu.L.: Computable functionals of finite type, Algebra
      and Logic 11 (1972), 203-247.

[8]   Ershov, Yu.L.: The theory of numerations, Vol 2 (in russian),
      Novosibirsk (1973).

[9]   Ershov, Yu.L.: Maximal and everywhere defined functionals,
      Algebra and Logic 13 (1974), 210-225.

[10] Feferman, S.: Inductive schemata and recursively continuous
      functionals, in R.O. Gandy and J.M.E. Hyland (eds.), Logic
      Colloquium '76, North-Holland (1977), 373-392.

[11] Fenstad, J.E.: General Recursion Theory, Springer Verlag 1980.

[12] Friedman, H.: Algorithmic procedures, generalized Turing algo-
      rithms, and elementary recursion theory, in R.O. Gandy and
      C.E.M. Yates (eds.), Logic Colloquium '69, North-Holland (1971),
      316 - 389.

[13] Gandy, R.O. and Hyland, J.M.E.: Computable and recursively
      countable functions of higher type, in R.O. Gandy and J.M.E.
      Hyland (eds.), Logic Colloquium '76, North-Holland (1977),407-438.

[14] Grilliot, T.: On effectively discontinuous type-2 objects,
      J.S.L. 36 (1971), 245-248.

[15] Harrington, L.: Contribution to recursion theory in higher types,
      Ph.D. Thesis, MIT 1973.

[16] Harrington, L. and MacQueen, D.B.: Selection in abstract recursion theory, J.S.L. 41 (1976), 153-158.

[17] Hinman, P.G.: Degrees of continuous functionals, J.S.L. 38 (1973), 393-395.

[18] Hinman, P.G.: Recursion-Theoretic Hierarchies, Springer Verlag 1978.

[19] Hyland, J.M.E.: Recursion on the countable functionals, Ph.D. Thesis, Oxford 1975.

[20] Hyland, J.M.E.: Filter spaces and continuous functionals, Ann. Math. Log. 16 (1979), 101-143.

[21] Hyland, J.M.E.: The intrinsic recursion theory on the countable or continuous functionals, in J.E. Fenstad, R.O. Gandy and G.E. Sacks (eds.): Generalized Recursion Theory II, North-Holland 1978, 135-145.

[22] Kleene, S.C.: Recursive functionals and Quantifiers of finite types I, T.A.M.S. 91 (1959), 1-52; and II (1963), 106-142.

[23] Kleene, S.C.: Countable functionals, in A. Heyting (ed.) Constructivity in mathematics, North-Holland (1959), 81-100.

[24] Kreisel, G.: Interpretation of analysis by means of functionals of finite type, in A. Heyting (ed.) Constructivity in mathematics North-Holland (1959), 101-128.

[25] Kuratowski, C.: Topologie Vol I, Warsawa 1952.

[26] MacQueen, D.B.: Post's problem for recursion in higher types, Ph.D. Thesis, M.I.T. 1972.

[27] Moldestad, J.: Computations in Higher Types, Lecture Notes in Mathematics, No 574, Springer Verlag 1977.

[28] Moldestad, J. and Normann, D.: Models for recursion theory, J.S.L. 47 (1976) 719-729.

[29] Moldestad, J., Stoltenberg-Hansen, V. and Tucker, J.V.: Finite algoritmic Procedures and inductive Definability, Oslo Preprint Series, No 6, 1978.

[30] Moldestad, J., Stoltenberg-Hansen, V. and Tucker, J.V.: Finite algoritmic procedures and computation theories, Oslo Preprint Series, No 7, 1978.

[31] Moschovakis, Y.N.: Hyperanalytic predicates, T.A.M.S. 129 (1967) 249-282.

[32] Normann, D.: A continuous type-2 functional with non-collapsing hierarchy, J.S.L. 43 (1978), 487-491.

As a preprint issue: On a problem of S. Wainer, Oslo Preprint Series No 13, 1976.

[33] Normann, D.: Set recursion,in J.E. Fenstad, R.O. Gandy and G.E. Sacks (eds.):Generalized Recursion Theory II, North-Holland 1978, 303-320.

[34] Normann, D.: Countable functionals and the analytic hierarchy, Oslo Preprint Series No 17, 1977.

[35] Normann, D. and Wainer, S.S.: The 1-section of a countable functional, Oslo Preprint Series No 4, 1978, to appear in J.S.L.

[36] Normann, D.: A classification of higher type functionals in F.V. Jensen, B.H. Mayoh and K.K. Møller (eds.) Proceedings from the 5th Scandinavian Logic Symposium, Aalborg Univ. Press 1979, 301-308

[37] Platek, R.A.: Foundations of recursion theory, Thesis, Stanford University 1966.

[38] Rogers Jr, H.: Theory of Recursive Functions and Effective Computability, McGraw-Hill 1967.

[39] Sacks, G.E.: Degrees of Unsolvability, Annals of Mathematical Studies, Princeton University Press, 1963.

[40] Sacks, G.E.: The 1-section of a type-n object, in J.E. Fenstad and P.G. Hinman (eds.) Generalized Recursion Theory, North-Holland 1974, 81-93.

[41] Sacks, G.E.: The k-section of a type-n object, Americal Journal of Mathematics 99 (1977), 901-917.

[42] Sacks, G.E.: Higher Recursion Theory, Monograph, to appear.

[43] Shoenfield, J.R.: Degrees of Unsolvability, Mathematics Studies 2, North-Holland/American Elsevier 1971.

[44] Tucker, J.V.: Computing in algebraic systems, Oslo Preprint Series No 12, 1978.

[45] Wainer, S.S.: The 1-section of a non-normal type-2 object, in J.E. Fenstad, R.O. Gandy and G.E. Sacks (eds.) Generalized Recursion Theory II, North-Holland 1978, 407-417.

[46] Wainer,S.S.: A hierarchy for the 1-section of any type two object, Journal of Symbolic Logic 39 (1974) 89-94.

ALPHABETIC LIST OF CONCEPTS

LIST OF SYMBOLS

Vol. 700: Module Theory, Proceedings, 1977. Edited by C. Faith and S. Wiegand. X, 239 pages. 1979.

Vol. 701: Functional Analysis Methods in Numerical Analysis, Proceedings, 1977. Edited by M. Zuhair Nashed. VII, 333 pages. 1979.

Vol. 702: Yuri N. Bibikov, Local Theory of Nonlinear Analytic Ordinary Differential Equations. IX, 147 pages. 1979.

Vol. 703: Equadiff IV, Proceedings, 1977. Edited by J. Fábera. XIX, 441 pages. 1979.

Vol. 704: Computing Methods in Applied Sciences and Engineering, 1977, I. Proceedings, 1977. Edited by R. Glowinski and J. L. Lions. VI, 391 pages. 1979.

Vol. 705: O. Forster und K. Knorr, Konstruktion verseller Familien kompakter komplexer Räume. VII, 141 Seiten. 1979.

Vol. 706: Probability Measures on Groups, Proceedings, 1978. Edited by H. Heyer. XIII, 348 pages. 1979.

Vol. 707: R. Zielke, Discontinuous Čebyšev Systems. VI, 111 pages. 1979.

Vol. 708: J. P. Jouanolou, Equations de Pfaff algébriques. V, 255 pages. 1979.

Vol. 709: Probability in Banach Spaces II. Proceedings, 1978. Edited by A. Beck. V, 205 pages. 1979.

Vol. 710: Séminaire Bourbaki vol. 1977/78, Exposés 507–524. IV, 328 pages. 1979.

Vol. 711: Asymptotic Analysis. Edited by F. Verhulst. V, 240 pages. 1979.

Vol. 712: Equations Différentielles et Systèmes de Pfaff dans le Champ Complexe. Edité par R. Gérard et J.-P. Ramis. V, 364 pages. 1979.

Vol. 713: Séminaire de Théorie du Potentiel, Paris No. 4. Edité par F. Hirsch et G. Mokobodzki. VII, 281 pages. 1979.

Vol. 714: J. Jacod, Calcul Stochastique et Problèmes de Martingales. X, 539 pages. 1979.

Vol. 715: Inder Bir S. Passi, Group Rings and Their Augmentation Ideals. VI, 137 pages. 1979.

Vol. 716: M. A. Scheunert, The Theory of Lie Superalgebras. X, 271 pages. 1979.

Vol. 717: Grosser, Bidualräume und Vervollständigungen von Banachmoduln. III, 209 pages. 1979.

Vol. 718: J. Ferrante and C. W. Rackoff, The Computational Complexity of Logical Theories. X, 243 pages. 1979.

Vol. 719: Categorial Topology, Proceedings, 1978. Edited by H. Herrlich and G. Preuß. XII, 420 pages. 1979.

Vol. 720: E. Dubinsky, The Structure of Nuclear Fréchet Spaces. V, 187 pages. 1979.

Vol. 721: Séminaire de Probabilités XIII. Proceedings, Strasbourg, 1977/78. Edité par C. Dellacherie, P. A. Meyer et M. Weil. VII, 647 pages. 1979.

Vol. 722: Topology of Low-Dimensional Manifolds. Proceedings, 1977. Edited by R. Fenn. VI, 154 pages. 1979.

Vol. 723: W. Brandal, Commutative Rings whose Finitely Generated Modules Decompose. II, 116 pages. 1979.

Vol. 724: D. Griffeath, Additive and Cancellative Interacting Particle Systems. V, 108 pages. 1979.

Vol. 725: Algèbres d'Opérateurs. Proceedings, 1978. Edité par P. de la Harpe. VII, 309 pages. 1979.

Vol. 726: Y.-C. Wong, Schwartz Spaces, Nuclear Spaces and Tensor Products. VI, 418 pages. 1979.

Vol. 727: Y. Saito, Spectral Representations for Schrödinger Operators With Long-Range Potentials. V, 149 pages. 1979.

Vol. 728: Non-Commutative Harmonic Analysis. Proceedings, 1978. Edited by J. Carmona and M. Vergne. V, 244 pages. 1979.

Vol. 729: Ergodic Theory. Proceedings, 1978. Edited by M. Denker and K. Jacobs. XII, 209 pages. 1979.

Vol. 730: Functional Differential Equations and Approximation of Fixed Points. Proceedings, 1978. Edited by H.-O. Peitgen and H.-O. Walther. XV, 503 pages. 1979.

Vol. 731: Y. Nakagami and M. Takesaki, Duality for Crossed Products of von Neumann Algebras. IX, 139 pages. 1979.

Vol. 732: Algebraic Geometry. Proceedings, 1978. Edited by K. Lønsted. IV, 658 pages. 1979.

Vol. 733: F. Bloom, Modern Differential Geometric Techniques in the Theory of Continuous Distributions of Dislocations. XII, 206 pages. 1979.

Vol. 734: Ring Theory, Waterloo, 1978. Proceedings, 1978. Edited by D. Handelman and J. Lawrence. XI, 352 pages. 1979.

Vol. 735: B. Aupetit, Propriétés Spectrales des Algèbres de Banach. XII, 192 pages. 1979.

Vol. 736: E. Behrends, M-Structure and the Banach-Stone Theorem. X, 217 pages. 1979.

Vol. 737: Volterra Equations. Proceedings 1978. Edited by S.-O. Londen and O. J. Staffans. VIII, 314 pages. 1979.

Vol. 738: P. E. Conner, Differentiable Periodic Maps. 2nd edition, IV, 181 pages. 1979.

Vol. 739: Analyse Harmonique sur les Groupes de Lie II. Proceedings, 1976–78. Edited by P. Eymard et al. VI, 646 pages. 1979.

Vol. 740: Séminaire d'Algèbre Paul Dubreil. Proceedings, 1977–78. Edited by M.-P. Malliavin. V, 456 pages. 1979.

Vol. 741: Algebraic Topology, Waterloo 1978. Proceedings. Edited by P. Hoffman and V. Snaith. XI, 655 pages. 1979.

Vol. 742: K. Clancey, Seminormal Operators. VII, 125 pages. 1979.

Vol. 743: Romanian-Finnish Seminar on Complex Analysis. Proceedings, 1976. Edited by C. Andreian Cazacu et al. XVI, 713 pages. 1979.

Vol. 744: I. Reiner and K. W. Roggenkamp, Integral Representations. VIII, 275 pages. 1979.

Vol. 745: D. K. Haley, Equational Compactness in Rings. III, 167 pages. 1979.

Vol. 746: P. Hoffman, ι-Rings and Wreath Product Representations. V, 148 pages. 1979.

Vol. 747: Complex Analysis, Joensuu 1978. Proceedings, 1978. Edited by I. Laine, O. Lehto and T. Sorvali. XV, 450 pages. 1979.

Vol. 748: Combinatorial Mathematics VI. Proceedings, 1978. Edited by A. F. Horadam and W. D. Wallis. IX, 206 pages. 1979.

Vol. 749: V. Girault and P.-A. Raviart, Finite Element Approximation of the Navier-Stokes Equations. VII, 200 pages. 1979.

Vol. 750: J. C. Jantzen, Moduln mit einem höchsten Gewicht. III, 195 Seiten. 1979.

Vol. 751: Number Theory, Carbondale 1979. Proceedings. Edited by M. B. Nathanson. V, 342 pages. 1979.

Vol. 752: M. Barr, *-Autonomous Categories. VI, 140 pages. 1979.

Vol. 753: Applications of Sheaves. Proceedings, 1977. Edited by M. Fourman, C. Mulvey and D. Scott. XIV, 779 pages. 1979.

Vol. 754: O. A. Laudal, Formal Moduli of Algebraic Structures. III, 161 pages. 1979.

Vol. 755: Global Analysis. Proceedings, 1978. Edited by M. Grmela and J. E. Marsden. VII, 377 pages. 1979.

Vol. 756: H. O. Cordes, Elliptic Pseudo-Differential Operators – An Abstract Theory. IX, 331 pages. 1979.

Vol. 757: Smoothing Techniques for Curve Estimation. Proceedings, 1979. Edited by Th. Gasser and M. Rosenblatt. V, 245 pages. 1979.

Vol. 758: C. Năstăsescu and F. Van Oystaeyen; Graded and Filtered Rings and Modules. X, 148 pages. 1979.